IMO Problems, Theorems, and Methods

Combinatorics

Mathematical Olympiad Series

ISSN: 1793-8570

Series Editors: Lee Peng Yee *(Nanyang Technological University, Singapore)*
Xiong Bin *(East China Normal University, China)*

Published

The complete list of the published volumes in the series can be found at
http://www.worldscientific.com/series/mos

Vol. 26 | Mathematical Olympiad Series

IMO Problems, Theorems, and Methods

Combinatorics

Authors

Guangyu Xu
Zhenhua Qu
East China Normal University, China

Translators

Zhenhua Qu
Jinhua Chen
East China Normal University, China

Proofreader

Jiu Ding
School of Mathematics and Natural Sciences,
University of Southern Mississippi, USA

Copy Editors

Lingzhi Kong, Liyu Zhang, and Ming Ni
East China Normal University Press, China

East China Normal University Press

World Scientific

Published by

East China Normal University Press
3663 North Zhongshan Road
Shanghai 200062
China

and

World Scientific Publishing Co. Pte. Ltd.
5 Toh Tuck Link, Singapore 596224
USA office: 27 Warren Street, Suite 401-402, Hackensack, NJ 07601
UK office: 57 Shelton Street, Covent Garden, London WC2H 9HE

Library of Congress Control Number: 2025004967

British Library Cataloguing-in-Publication Data
A catalogue record for this book is available from the British Library.

Mathematical Olympiad Series — Vol. 26
IMO PROBLEMS, THEOREMS, AND METHODS
Combinatorics

ISBN 978-981-98-0336-1 (hardcover)
ISBN 978-981-98-0691-1 (paperback)
ISBN 978-981-98-0337-8 (ebook for institutions)
ISBN 978-981-98-0338-5 (ebook for individuals)

For any available supplementary material, please visit
https://www.worldscientific.com/worldscibooks/10.1142/14101#t=suppl

Desk Editors: Nambirajan Karuppiah/Angeline Husni

Typeset by Stallion Press
Email: enquiries@stallionpress.com

Preface

It is generally believed that formal mathematics competitions began with a contest held in Hungary in 1894, an event that gradually garnered attention worldwide. People aptly liken mathematics competitions to "Mental Gymnastics." In 1934, the Soviet Union straightforwardly termed it the "Mathematical Olympiad," a designation that reflects the Olympic spirit of pursuing excellence in intellect more vividly than the previous term, "mathematics competition."

By 1959, the internationalization of mathematics competitions had matured, leading to the inception of the "International Mathematical Olympiad" (IMO). The first IMO was held in Brasov, Romania in 1959. As of 2023, the IMO has successfully been held 64 times, except for 1980 when it was not conducted.

The IMO is typically held in July each year, and the format has become standardized: the official competition spans two days, with contestants tackling three problems in 4.5 hours each day, each problem worth a maximum of 7 points, totaling 42 points. Each participating team consists of six contestants, accompanied by a Leader and a Deputy Leader. Approximately half of the contestants receive medals, with about 1/12 of the contestants earning gold medals, 2/12 receiving silver medals, and 3/12 obtaining bronze medals.

The IMO is currently one of the most influential secondary school mathematics competitions worldwide. In recent years, over 100 countries and regions have participated in this event, including all major nations globally.

Problems for the IMO are submitted by the participating teams and then reviewed and selected by a problem selection committee organized by the host country. This committee narrows down the submissions to approximately 30 Shortlist problems, covering algebra, geometry, number theory, and combinatorics, with about seven to eight problems on each topic. These are then presented to the Jury Meeting, composed of team leaders, who discuss and vote to decide on the six problems that will constitute the official competition paper. The host country does not provide any problems.

This event has played a significant role in promoting the exchange of mathematical education among nations, enhancing the level of mathematical education, facilitating mutual learning and understanding among young students worldwide, stimulating a broad interest in mathematics among secondary school students, and identifying and nurturing mathematically gifted students.

The development over more than 60 years is the result of the collective efforts of mathematicians, organizers, and contestants, and is worthy of reflection and study. Particularly deserving of study are the evolution of the competition problems, the mathematical ideas, and methods involved. Indeed, several colleagues from the International Mathematical Olympiad Research Center at East China Normal University had envisioned research and publication before the 60th IMO. For this purpose, we initiated several seminars involving over 10 people. For special reasons, this work was delayed. Based on the mathematical domains covered by the IMO problems — algebra, geometry, number theory, and combinatorics — we planned to compile the work into four volumes, with the general title *IMO Problems, Theorems, and Methods*, to be included in the "IMO Study Series."

Each volume begins with an introduction that provides an overview of the IMO. Subsequent chapters introduce relevant foundational knowledge and methods, followed by a reclassification and organization of past IMO problems. For some problems, multiple solutions are provided, along with a difficulty analysis. It is worth noting that some problems do not fit neatly into a single topic, as they may involve both algebra and number theory, or algebra and combinatorics. We primarily categorize them based on the topic under which they were placed on the Shortlist.

The four volumes titled *IMO Problems, Theorems, and Methods* were conceived with an overall writing plan proposed by myself, with the authors collectively discussing and refining the plan. The majority of the initial drafts were completed by Jinhua Chen (Algebra), Tianqi Lin (Geometry), Gengyu Zhang (Number Theory), and Guangyu Xu (Combinatorics).

The first three volumes were supplemented, consolidated, and finalized by myself, while the combinatorics volume was supplemented, consolidated, and finalized by Zhenhua Qu.

We extend our gratitude to the leaders and contestants of the Chinese IMO teams over the years, as some elegant solutions included in the book were contributed by them. During the compilation of this book, we consulted various domestic and international sources, which are too numerous to acknowledge individually here.

While the authors have diligently studied the IMO problems and provided thoughtful strategies and solutions, errors and inaccuracies may occur due to our limitations. We sincerely invite readers to offer corrections and feedback.

The translation of the algebra volume in this series was done by Jinhua Chen and Bin Xiong; the geometry volume was translated by Xinyuan Yang; the number theory volume was translated by Gengyu Zhang; and the combinatorics volume was translated by Zhenhua Qu and Jinhua Chen. Jiu Ding revised the translations of all four books.

Bin Xiong
June 2024

About the Authors

Guangyu Xu is a Ph.D. student from School of Mathematical Sciences, East China Normal University. He completed his undergraduate studies in Mathematics and Applied Mathematics from Tsinghua University. During his doctoral studies, his research focuses on mathematics education and mathematics competitions, with a particular emphasis on combinatorial mathematics. He won a silver medal during the 29th Chinese Mathematical Olympiad and has maintained a strong passion for mathematics competitions over the years. He has also participated in the proposition and grading work for various major mathematics competitions in China multiple times.

Zhenhua Qu is an associate professor of the School of Mathematical Sciences, East China Normal University. He received his PHD from the University of Texas at Austin, USA. His research areas include algebraic geometry, number theory and mathematics education. He is actively involved in national mathematical competitions and has proposed a great number of problems. He was the leader of Chinese IMO team in 2018, and deputy leader in 2021–2023.

About the Translators

Zhenhua Qu is an associate professor of the School of Mathematical Sciences, East China Normal University. He received his PHD from the University of Texas at Austin, USA. His research areas include algebraic geometry, number theory and mathematics education. He is actively involved in national mathematical competitions and has proposed a great number of problems. He was the leader of Chinese IMO team in 2018, and deputy leader in 2021–2023.

Jinhua Chen is a doctoral candidate in Mathematical Education at East China Normal University, with research interests in mathematics competitions and gifted education. He serves as an external mentor for mathematics competition courses at several high schools. Chen attended the Affiliated High School of South China Normal University and Guangdong Olympic School during his secondary education. He has excelled in various competitions, including the Chinese High School Mathematics League, the American Mathematics Competition 12, the Chinese College Mathematics Competition, and the COMAP Mathematical/Interdisciplinary Contest in Modeling.

Contents

Introduction to the IMO

The International Mathematical Olympiad (IMO), established in the year 1959, represents one of the foremost intellectual endeavors at the highest tier for youth on a global scale. Prior to 1959, numerous countries around the world had already initiated the organization of mathematics competitions, thereby laying the groundwork for the inception of the IMO.

In 1891, the renowned physicist and President of the Hungarian Academy of Sciences, Loránd Eötvös (also known as Roland Eötvös), founded the Hungarian Mathematical and Physical Society. In 1894, he assumed the position of Minister of Education, and under his enthusiastic support, the Hungarian Mathematical and Physical Society initiated secondary school mathematics competitions. This competition, also known as the Eötvös Competition, offered winners the Eötvös Prize and the opportunity to pursue higher education. Subsequently, the Eötvös Competition was not held during 1919–1921 and 1944–1946 due to world political events. In 1947, under the leadership of János Surányi, the Eötvös Competition was reinstated and renamed the Kürschák Competition (named after József Kürschák). This competition has played a significant role in Hungary in nurturing numerous mathematicians and scientists, including Győző Zemplén, Lipót Fejér, Theodore von Kármán, Alfréd Haar, Dénes Kőnig, Marcel Riesz, Gábor Szegő, Tibor Radó, Edward Teller, and Tibor Szele. Interestingly, George Pólya also participated in the competition, but did not hand in his paper.

With the aim of identifying and nurturing mathematical talents, prominent mathematicians such as Boris Delaunay (also known as Delone), Grigorii Fikhtengol'ts, Dmitry Faddeev, and others organized the inaugural

Leningrad Mathematical Olympiad (LMO) in 1934, under the initiative of Boris Delaunay. Winners of this competition were granted the privilege of direct admission to the Mathematics Department of Leningrad State University without the need for entrance examinations. Following the example set by the LMO, in 1935, renowned mathematicians Pavel Aleksandrov and Andrey Kolmogorov, alongside the entire faculty of the Mathematics Department at Moscow State University, organized the first Moscow Mathematical Olympiad (MMO).

Subsequently, various regions throughout the Soviet Union started hosting their own Mathematical Olympiads, ultimately laying the foundation for the All-Russian Mathematical Olympiad, which was first conducted in 1961. In 1967, the responsibility for organizing the All-Russian Mathematical Olympiad was assumed by the Ministry of Education of the Soviet Union, leading to a renaming of the All-Russian Mathematical Olympiad as the All-Soviet-Union Mathematical Olympiad.

In fact, almost all the best mathematicians born in the Soviet Union after 1930 had participated in Mathematical Olympiads, usually achieving first prizes. This distinguished group includes Fields Medal awardees such as Sergei Novikov, Grigory Margulis, Vladimir Drinfeld, Maxim Kontsevich, Grigori Perelman, and Stanislav Smirnov. Although having claimed that he was never particularly interested in Mathematical Olympiads, Sergei Novikov did secure a second prize in the MMO when he was in eighth grade.

While the United States of America Mathematical Olympiad (USAMO) was first held in 1972, the United States had a longstanding tradition of organizing mathematics competitions prior to that. In 1921, William Lowell Putnam published an article in the *Harvard Graduates' Magazine*, proposing the idea of conducting a university-level mathematics team competition. Following his passing, the Putnam family established the William Lowell Putnam Intercollegiate Memorial Fund to support the organization of the William Lowell Putnam Mathematical Competition (Putnam Competition), administered by the Mathematical Association of America.

With the assistance of George David Birkhoff, the first Putnam Competition took place in 1938, and it has been held annually since then; the top five ranking participants are designated as Putnam Fellows. Due to wartime conditions, the competition was not held from 1943 to 1945. In 1946, George Pólya, Tibor Radó, and Irving Kaplansky (Putnam Fellow in 1938) formed the Putnam Competition Committee, thus reestablishing the competition, but the responsibility for administration was undertaken by

Garrett Birkhoff, the son of George David Birkhoff, and his colleagues in the Harvard University Department of Mathematics.

Many participants in the Putnam Competition have gone on to become prominent mathematicians and scientists. John Milnor, David Mumford, Daniel Quillen, Paul Cohen, John G. Thompson, and Manjul Bhargava have been recipients of Fields Medal. Richard Feynman, Kenneth Geddes Wilson, Steven Weinberg, and Murray Gell-Mann have received Nobel Prize in Physics, while John Nash was awarded the Nobel Prize in Economic Sciences. Additionally, numerous Putnam Fellows have been elected as members of the National Academy of Sciences in the United States.

Building upon the foundation of existing mathematics competitions in many countries, particularly in Eastern European nations, Romania proposed in 1956 the organization of an international mathematics competition involving seven Eastern European countries. This proposal led to the inaugural IMO held in 1959.

The first IMO was held in Braşov, Romania, in 1959. As of 2023, the IMO has been successfully held 64 times, except for the year 1980 when it was not held for specific reasons. Apart from the 61st IMO, which was postponed to September in 2020 due to the impact of the COVID-19 pandemic, the IMO typically takes place in July each year.

The IMO has emerged as the most influential secondary school mathematics competition at present. In recent years, the number of countries and regions participating in this event has exceeded 100.

1 Evolution of the IMO

The first IMO, held in 1959, saw the participation of only 52 contestants from seven countries, including Romania, Hungary, Czechoslovakia, Bulgaria, Poland, the Soviet Union, and the German Democratic Republic. Subsequently, new countries and regions gradually joined this prestigious event. By the 20th IMO, also hosted by Romania in 1978, approximately 20 countries and regions participated (with 21 in the 19th, 17 in the 20th, and 23 in the 21st). The number of participating contestants also reached 132. The historical participation trends are depicted in Figure 1.

As the influence of the IMO continued to expand, the number of participating countries and regions, as well as the number of contestants, grew rapidly. The most significant increase in the number of participating countries and regions occurred in the 34th IMO, which was held in Turkey in 1993. In comparison to the 33rd IMO held in Russia in 1992, the number

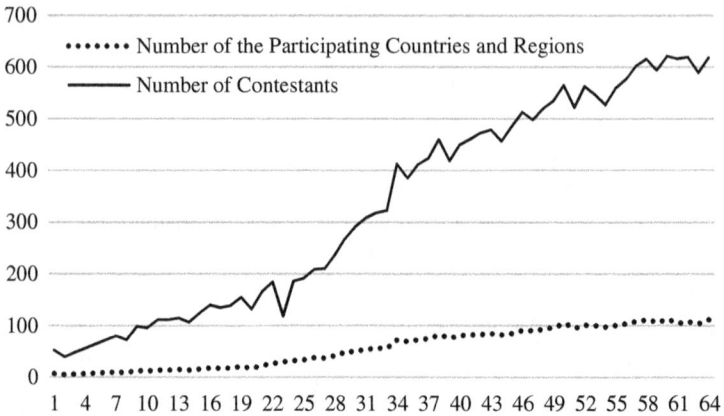

Figure 1 Numbers of Participating Countries and Regions, as Well as the Contestants, in the First 64 IMOs

increased by 17 countries and regions, reaching a total of 73, with 413 contestants. By the 40th IMO, which was still hosted by Romania in 1999, the number of participating countries and regions had reached 81, with 450 contestants.

The first instance of the number of participating countries and regions surpassing one hundred (104) occurred during the 50th IMO, held in Germany in 2009, with a total of 565 contestants. Among the first 64 IMOs, the biggest number of participating countries and regions, as well as the largest number of contestants, was observed in the 60th IMO, hosted by the United Kingdom in 2019, where 621 contestants from 112 countries and regions took part. In the 64th IMO held in Japan in 2023, there were 618 contestants from 112 countries and regions.

As evident from Appendix A, the IMO is primarily hosted by European countries. Moreover, as the number of participating countries and regions in the IMO has increased, it is no longer confined to the seven founding member countries, and many new participating countries and regions have also begun to organize the IMO.

2 Problems in the IMO

The IMO is scheduled to take place annually in July. Each participating country or region officially sends a delegation consisting of six contestants, along with one team leader and one deputy leader. The official competition spans two days, with each day featuring three problems to be solved within a four-and-a-half-hour timeframe. Each problem carries a maximum score of 7 points, resulting in a total maximum score of 42 points, while the total maximum score of the team is 252 points.

In early IMOs, the number of problems and their individual point values varied from one session to another. For instance, the 2nd and 4th IMOs featured seven problems, while all other IMOs had six problems each. Additionally, in the 13th IMO, although the total score remained at 42 points, the six problems were allocated point values of 5, 7, 9, 6, 7, and 8, respectively. It was only from the 22nd IMO, held in the United States in 1981, that the IMO problems have become standardized, with each problem carrying 7 points and a total of six problems.

The number of contestants in each delegation has also become stable at six individuals starting from the 24th IMO held in France in 1983.

2.1 The number of problems

The mathematical domains covered by IMO problems encompass four major topics: algebra, combinatorics, geometry, and number theory. These are also the primary focus in various national mathematics competitions.

Across the 1st–64th IMOs, a total of 386 problems have been featured, as shown in Table 1. Among them, geometry problems are the most numerous, with 123 problems, while number theory problems are the least, with 75 problems. Algebra problems account for approximately one-quarter of the total, comprising 101 problems.

Remarkably, in the first 64 IMOs, there were four sessions when three combinatorics problems were presented, specifically in the 30th, 31st, 52nd,

Table 1 Numbers of Problems with Different Topics in the First 64 IMOs

Session	Topic			
	Algebra	Combinatorics	Geometry	Number Theory
1–10	20	6	29	7
11–20	20	12	18	10
21–30	14	16	18	12
31–40	13	16	15	16
41–50	15	11	20	14
51–60	13	19	18	11
61–64	6	7	6	5
Total	101	87	123	75

and 61st IMOs. In 22 IMOs, two combinatorics problems were featured, while in 31 IMOs, only one combinatorics problem was included. In seven IMOs, combinatorics problem was absent.

2.2 *The difficulty level of problems*

Typically, the difficulty of a problem is correlated with its problem number in the IMO.

Starting from the 24th IMO, the point value of each problem and the number of contestants per team have become standardized. Therefore, an analysis of the average scores of the 246 problems in the 24th–64th IMOs is presented in Table 2. It can be observed that the first and fourth problems in each IMO are relatively easy, with average scores generally exceeding 3 points. The second and fifth problems are relatively challenging, with average scores mainly ranging from 1 to 4 points. The third and sixth problems are exceptionally difficult, with average scores generally falling below 2 points.

Table 2 Numbers of Problems with Different Average Scores in the 24th–64th IMOs

Problem Number	Problem Mean				
	0–1	1–2	2–3	3–4	4–7
Problem 1	0	0	5	13	23
Problem 2	2	6	18	9	6
Problem 3	22	12	4	3	0
Problem 4	0	0	7	17	17
Problem 5	4	10	14	10	3
Problem 6	24	11	6	0	0
Total	52	39	54	52	49

The 246 problems are categorized into four topics: algebra, combinatorics, geometry, and number theory, as shown in Table 3. Notably, there is a relatively large representation of combinatorics and geometry problems. Combining this information with Table 1, it is evident that in the first 23 IMOs, algebra and geometry problems were predominant.

Furthermore, in early IMOs, geometry problems predominantly appeared in the 1st/4th and 2nd/5th positions. However, starting from the 41st to 50th IMOs, geometry problems were more commonly found in the 1st/4th and 3rd/6th positions. Similarly, algebra problems were more frequent in the 1st/4th and 2nd/5th positions, combinatorics problems were more prevalent in the 2nd/5th and 3rd/6th positions, while the quantity of

Table 3 Numbers of Problems with Different Topics in the 24th–64th IMOs

	Topic										
	Algebra			Combinatorics			Geometry			Number Theory		
Session	1, 4	2, 5	3, 6	1, 4	2, 5	3, 6	1, 4	2, 5	3, 6	1, 4	2, 5	3, 6
24–30	4	1	5	4	3	5	5	7	0	1	3	4
31–40	4	5	4	5	4	7	6	9	0	5	2	9
41–50	3	8	4	3	2	6	10	6	5	5	4	5
51–60	5	6	2	4	7	8	7	3	6	3	4	4
61–64	1	3	2	2	2	3	3	1	2	2	2	1
Total	17	23	17	18	18	29	31	26	13	16	15	23
		57			65			70			54	

number theory problems across different problem numbers does not differ significantly.

From Table 4, it can be observed that among the four topics, the numbers of problems with an average score ranging from 2 to 4 points are quite similar. However, in the combinatorics topic, there is a higher quantity of challenging problems, with 31 problems having an average score between 0 and 2 points. Conversely, the geometry topic has the largest number of relatively easy problems, with 23 problems scoring above 4 points. This discrepancy is largely due to the fact that there are 29 combinatorics problems in the 3rd/6th positions, and 31 geometry problems in the 1st/4th positions.

Table 4 Numbers of Problems with Different Average Scores by Topics in the 24th–64th IMOs

	Problem Mean					
Topic	0–1	1–2	2–3	3–4	4–7	Total
Algebra	7	14	19	6	11	57
Combinatorics	20	11	12	14	8	65
Geometry	13	5	14	15	23	70
Number theory	12	9	9	17	7	54
Total	52	39	54	52	49	246

Furthermore, when considering Table 3 and Table 4, it becomes apparent that among the four topics, the number of problems with an average score ranging from 0 to 2 points closely aligns with the number of problems in the 3rd/6th positions. There are slightly more problems with an average score between 2 to 4 points compared to those in the 2nd/5th positions, and slightly fewer problems with an average score exceeding 4 points compared to those in the 1st/4th positions. This indicates that even the seemingly easier problems in the IMO are not as straightforward as they might appear.

Notably, among these 246 problems, the lowest average score is attributed to IMO 58-3 (Combinatorics, proposed by Austria):

A hunter and an invisible rabbit play a game in a Euclidean plane. The rabbit's starting point, A_0, and the hunter's starting point, B_0, are the same. After $n - 1$ rounds of the game, the rabbit is at point A_{n-1} and the hunter is at point B_{n-1}. In the nth round of the game, three things occur in order.

(i) The rabbit moves invisibly to a point A_n such that the distance between A_{n-1} and A_n is exactly 1.

(ii) A tracking device reports a point P_n to the hunter. The only guarantee provided by the tracking device to the hunter is that the distance between P_n and A_n is at most 1.

(iii) The hunter moves visibly to a point B_n such that the distance between B_{n-1} and B_n is exactly 1.

Is it always possible, no matter how the rabbit moves, and no matter what points are reported by the tracking device, for the hunter to choose her moves so that after 10^9 rounds she can ensure that the distance between her and the rabbit is at most 100?

This unconventional problem received an average score of only 0.042 points. Only two contestants, Mikhail Ivanov from Russia and Linus Cooper from Australia, achieved a perfect score of 7 points. Joe Benton from the United Kingdom scored 5 points, Pavel Hudec from Czech Republic earned 4 points, Hadyn Ka Ming Tang from Australia, Yahor Dubovik from Belarus, and Jeonghyun Ahn from South Korea each scored 1 point.

Furthermore, among the 20 lowest scoring problems in the 24th–64th IMOs, nearly all of them appeared in the 3rd/6th positions, as indicated in Table 5. There were three algebra problems, eight combinatorics problems, six geometry problems, and three number theory problems among them. These eight combinatorics problems were:

• **(IMO 48-3, proposed by Russia).** In a mathematical competition some competitors are friends. Friendship is always mutual. Call a group of competitors a *clique* if each two of them are friends. (In particular, any group of fewer than two competitors is a clique.) The number of members of a clique is called its *size*.

Table 5 The 20 Problems with the Lowest Average Scores in the 24th–64th IMOs

Problem	Mean	Topic	Problem	Mean	Topic
IMO 58-3	0.042	Combinatorics	IMO 54-6	0.296	Combinatorics
IMO 48-6	0.152	Algebra	IMO 48-3	0.304	Combinatorics
IMO 50-6	0.168	Combinatorics	IMO 52-6	0.318	Geometry
IMO 47-6	0.187	Geometry	IMO 53-6	0.336	Number Theory
IMO 57-3	0.251	Number Theory	IMO 55-6	0.339	Combinatorics
IMO 49-6	0.260	Geometry	IMO 56-6	0.355	Combinatorics
IMO 64-6	0.275	Geometry	IMO 51-6	0.368	Algebra
IMO 59-3	0.278	Combinatorics	IMO 62-3	0.372	Geometry
IMO 61-6	0.282	Combinatorics	IMO 62-2	0.375	Algebra
IMO 58-6	0.294	Number Theory	IMO 60-6	0.403	Geometry

Given that, in this competition, the largest size of a clique is even, prove that the competitors can be arranged in two rooms such that the largest size of a clique contained in one room is the same as the largest size of a clique contained in the other room.

- **(IMO 50-6, proposed by Russia).** Let $a_1, a_2, \ldots, a_n$ be distinct positive integers and let M be a set of $n - 1$ positive integers not containing $s = a_1 + a_2 + \cdots + a_n$. A grasshopper is to jump along the real axis, starting at the point 0 and making n jumps to the right with lengths $a_1, a_2, \ldots, a_n$ in some order. Prove that the order can be chosen in such a way that the grasshopper never lands on any point in M.

- **(IMO 54-6, proposed by Russia).** Let $n \geq 3$ be an integer and consider a circle with $n + 1$ equally spaced points marked on it. Consider all labellings of these points with the numbers $0, 1, \ldots, n$ such that each label is used exactly once; two such labellings are the same if one can be obtained from the other by a rotation of the circle. A labelling is called *beautiful* if, for any four labels $a < b < c < d$ with $a + d = b + c$, the chord joining the points labelled a and d does not intersect the chord joining the points labelled b and c.

  Let M be the number of beautiful labellings, and let N be the number of ordered pairs (x, y) of positive integers such that $x + y \leq n$ and $\gcd(x, y) = 1$. Prove that $M = N + 1$.

- **(IMO 55-6, proposed by Austria).** A set of lines in a plane is in *general position* if no two are parallel and no three pass through the same point. A set of lines in general position cuts the plane into regions, some of which have a finite area; we call these its *finite regions*. Prove that for all sufficiently large n, in any set of n lines in general position it is possible to colour at least $\sqrt{n}$ of the lines blue in such a way that none of its finite regions has a completely blue boundary.

Note. Results with $\sqrt{n}$ replaced by $c\sqrt{n}$ will be awarded points depending on the value of the constant c.

• **(IMO 56-6, proposed by Australia).** A sequence $a_1, a_2, \ldots$ of integers satisfies the following conditions:

(i) $1 \le a_j \le 2015$ for all $j \ge 1$;
(ii) $k + a_k \ne l + a_l$ for all $1 \le k < l$.

Prove that there exist two positive integers b and N such that

$$\left| \sum_{j=m+1}^{n} (a_j - b) \right| \le 1007^2$$

for all integers m and n satisfying $n > m \ge N$.

• **(IMO 58-3, proposed by Austria).**

• **(IMO 59-3, proposed by Iran).** An *anti-Pascal triangle* is an equilateral triangular array of numbers such that, except for the numbers in the bottom row, each number is the absolute value of the difference of the two numbers immediately below it. For example, the following array is an anti-Pascal triangle with four rows which contains every integer from 1 to 10, as shown in Figure 2.

$$4$$
$$2 \quad 6$$
$$5 \quad 7 \quad 1$$
$$8 \quad 3 \quad 10 \quad 9$$

Figure 2 An Anti-Pascal Triangle

Does there exist an anti-Pascal triangle with 2018 rows which contains every integer from 1 to $1 + 2 + \cdots + 2018$?

• **(IMO 61-6, proposed by Chinese Taiwan).** Prove that there exists a positive constant c such that the following statement is true:

Consider an integer $n > 1$, and a set S of n points in a plane such that the distance between any two different points in S is at least 1. It follows that there is a line l separating S such that the distance from any point of S to l is at least $cn^{-\frac{1}{3}}$.

(A line l separates a set of points S if some segment joining two points in S crosses l.)

Note: Weaker results with $cn^{-\frac{1}{3}}$ replaced by $cn^{-\alpha}$ may be awarded points depending on the value of the constant $\alpha > \frac{1}{3}$.

2.3 The classification of problems

In the 1st–64th IMOs, there were 87 combinatorics problems, which can be categorized into six specialized subjects: enumerative combinatorics problems, existence problems, extremal combinatorial problems, operation and logical reasoning problems, combinatorial geometry problems, and graph theory problems.

There were 101 algebra problems, which can be categorized into five specialized subjects: equation problems, function problems, sequence problems, inequality problems, and other algebra problems.

There were 123 geometry problems, which can be categorized into seven specialized subjects: similarity and congruence problems, circle problems, power-of-point problems, special point and line problems, trigonometry problems, solid geometry problems, and geometric inequality problems.

There were 75 number theory problems, which can be categorized into three specialized subjects: divisibility-of-integer problems, modular arithmetic problems, and indeterminate equation problems.

For the 87 combinatorics problems categorized and analyzed, as shown in Table 6, it can be observed that the subject of combinatorial geometry problems had the largest number of problems, with 27 problems.

Table 6 Numbers of Combinatorics Problems in the First 64 IMOs

	Subject					
Session	Enumerative Combinatorics Problems	Existence Problems	Extremal Combinatorial Problems	Operation and Logical Reasoning Problems	Combinatorial Geometry Problems	Graph Theory Problems
1–10	2	0	0	2	2	1
11–20	1	3	2	0	5	0
21–30	7	2	0	1	5	1
31–40	2	1	2	4	4	3
41–50	4	3	0	2	2	0
51–60	3	3	0	5	7	1
61–64	0	1	0	2	2	2
Total	19	13	4	16	27	8

In the first 20 IMOs, combinatorics problems primarily focused on combinatorial geometry problems.

Subsequently, the numbers of enumerative combinatorics problems and operation and logical reasoning problems increased rapidly in the 21st–50th IMOs, with 13 and seven problems, respectively, while combinatorial geometry problems maintained a certain frequency.

In the 51st–64th IMOs, combinatorics problems were primarily centered around operation and logical reasoning problems and combinatorial geometry problems.

In the first 64 IMOs, there had been a total of 19 enumerative combinatorics problems, accounting for approximately 21.8% of all combinatorics problems. There were a total of 14 enumerative combinatorics problems in the 24th–64th IMOs, and as shown in Table 7, these problems frequently appeared as the 1st/4th problem.

Table 7   Numbers of Enumerative Combinatorics Problems in the 24th–64th IMOs

Enumerative Combinatorics Problem	Problem Number			Number of Problems in the First 64 IMOs
	1, 4	2, 5	3, 6	
Enumerative combinatorics problems	6	4	4	19

In the first 64 IMOs, there had been a total of 13 existence problems, accounting for approximately 14.9% of all combinatorics problems. There were a total of 10 existence problems in the 24th–64th IMOs, and as shown in Table 8, these problems frequently appeared as the 3rd/6th problem.

Table 8   Numbers of Existence Problems in the 24th–64th IMOs

Existence problem	Problem Number			Number of Problems in the First 64 IMOs
	1, 4	2, 5	3, 6	
Existence problems	2	3	5	13

In the first 64 IMOs, there had been a total of four extremal combinatorial problems, accounting for approximately 4.6% of all combinatorics problems.

In the first 64 IMOs, there had been a total of 16 operation and logical reasoning problems, accounting for approximately 18.4% of all combinatorics problems. These problems can be primarily categorized into three types: (1) logical reasoning problems, totaling two problems; (2) single-person operation problems, totaling nine problems; (3) double-person operation problems, totaling five problems.

There were a total of 14 operation and logical reasoning problems in the 24th–64th IMOs, and as shown in Table 9, these problems frequently appeared as the 3rd/6th problem, primarily focusing on single-person operation problems.

Table 9 Numbers of Operation and Logical Reasoning Problems in the 24th–64th IMOs

Operation and Logical Reasoning Problem	Problem Number			Number of Problems in the First 64 IMOs
	1, 4	2, 5	3, 6	
Logical reasoning problems	0	0	0	2
Single-person operation problems	3	2	4	9
Double-person operation problems	1	2	2	5
Total	4	4	6	16

In the first 64 IMOs, there had been a total of 27 combinatorial geometry problems, accounting for approximately 31.0% of all combinatorics problems. These problems can be primarily categorized into four types: (1) point and line problems, totaling 17 problems; (2) shape problems, totaling three problems; (3) covering, embedding, partitioning and patching problems, totaling two problems; (4) lattice point and grid problems, totaling five problems.

There were a total of 19 combinatorial geometry problems in the 24th–64th IMOs, and as shown in Table 10, these problems frequently appeared as the 3rd/6th problem. Most of these problems, totaling 11, were of the point and line problems. The other three types of combinatorial geometry problems were less frequent.

In the first 64 IMOs, there had been a total of eight graph theory problems, accounting for approximately 9.2% of all combinatorics problems. There were a total of six graph theory problems in the 24th–64th IMOs, and as shown in Table 11, these problems frequently appeared as the 3rd/6th problem.

Table 10 Numbers of Combinatorial Geometry Problems in the 24th–64th IMOs

Combinatorial Geometry Problem	Problem Number			Number of Problems in the First 64 IMOs
	1, 4	2, 5	3, 6	
Point and line problems	3	3	5	17
Shape problems	0	1	1	3
Covering, embedding, partitioning, and patching problems	0	0	1	2
Lattice point and grid problems	1	2	2	5
Total	4	6	9	27

Table 11 Numbers of Graph Theory Problems in the24th–64th IMOs

Graph Theory Problems	Problem Number			Numbers of Problems in the First 64 IMOs
	1, 4	2, 5	3, 6	
Graph theory problems	2	1	3	8

As shown in Table 12, among the combinatorics problems in the 24th–64th IMOs, the largest proportion of problems is attributed to combinatorial geometry problems. The average scores of combinatorial geometry problems are primarily concentrated between 0 and 2 points, which suggests that combinatorial geometry problems tend to be more challenging.

Table 12 Numbers of Combinatorics Problems with Different Average Scores in the 24th–64th IMOs

Combinatorics Problem	Problem Mean					Total
	0–1	1–2	2–3	3–4	4–7	
Enumerative combinatorics problems	3	1	3	5	2	14
Existence problems	5	1	3	1	0	10
Extremal combinatorial problems	0	1	1	0	0	2
Operation and logical reasoning problems	5	2	1	4	2	14
Combinatorial geometry problems	5	5	4	1	4	19
Graph theory problems	2	1	0	3	0	6
Total	20	11	12	14	8	65
		31		26	8	

Furthermore, the number of combinatorics problems in the 3rd/6th positions is roughly equal to the number of combinatorics problems with average scores of 0–2. However, the numbers of combinatorics problems in the 1st/4th and 2nd/5th positions differ from the numbers of combinatorics problems with average scores of 4–7 and 2–4, which implies that even the easier problems are not necessarily easy.

2.4 *The proposal for problems*

Problems in the IMO are proposed by the participating countries and regions, except the host. Usually, the team leaders are in charge of submitting problems with a limit of six, and these problems are then subjected to the selection by a selection committee composed of experts organized by the host. Approximately 30 problems are chosen as shortlist problems, with around eight problems in each of the four topics: algebra, geometry, combinatorics, and number theory. Subsequently, these problems are submitted to the Jury, which is comprised of team leaders from each participating country or region. The problems are discussed and voted upon to select the official examination problems. Once the problems are finalized, they are translated into five working languages: English, French, German, Russian, and Spanish. Each leader then translates the problems into their respective national languages, and contestants can choose from two languages in which to answer the problems.

Among the first 64 IMOs, combinatorics problems were contributed by 39 different countries and regions. The Netherlands had the biggest number of problems proposed, with a total of six. Both Hungary and Germany proposed five problems, while the United States, the Soviet Union, Russia, and Australia contributed four problems each. These seven countries collectively provided 32 problems. Remarkably, only the German Democratic Republic (in the 27th IMO in 1986) has proposed two combinatorics problems in one IMO session.

3 Awards in the IMO

In addition to selecting problems, the Jury has several other responsibilities, including establishing grading criteria, resolving discrepancies in grading between leaders and coordinators, and determining the number of gold, silver, and bronze medals, as well as the score thresholds. In each IMO, approximately 1/12 of the contestants receive a gold medal, 2/12 receive silver, and 3/12 receive bronze.

Apart from the gold, silver, and bronze medals, contestants who do not receive medals but attain a score of 7 on at least one problem in the IMO will receive an Honorable Mention. Contestants who deliver exceptionally elegant solutions to specific problems in the IMO will receive a Special Prize.

As depicted in Figure 3, starting from the 24th IMO, the cutoff scores for gold, silver, and bronze medals have gradually stabilized. The gold medal cutoff is approximately 29 points, the silver medal cutoff is around 22 points, and the bronze medal cutoff is roughly 15 points. Furthermore, the average score of all contestants closely aligns with the bronze medal cutoff. This indicates that the problem difficulty is well-balanced.

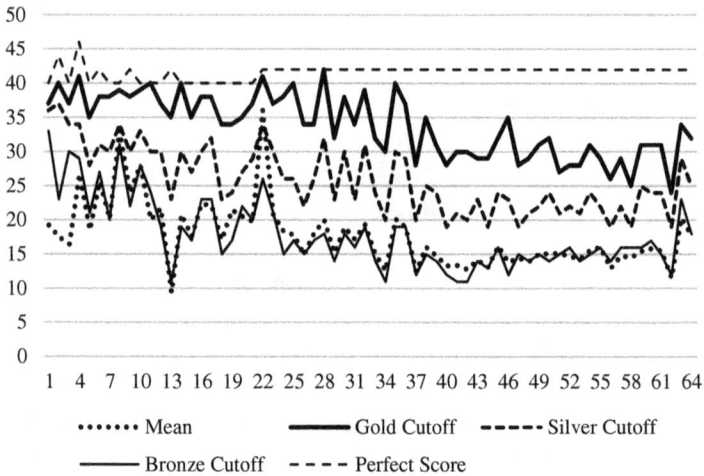

Figure 3   Medal Cutoff Scores in the First 64 IMOs

Interestingly, in the first 64 IMOs, there were three occasions where the gold medal cutoff was a perfect score, meaning only those who scored full marks could earn a gold medal. These three IMOs were: the 11th IMO (1969, Romania) with a perfect score of 40 points and three gold medalists; the 14th IMO (1972, Poland) with a perfect score of 40 points and eight gold medalists; and the 28th IMO (1988, Cuba) with a perfect score of 42 points and 22 gold medalists.

3.1 *Participation*

In the first 64 IMOs, a total of 269 contestants took part in four or more IMOs. Among them, two contestants attended seven IMOs, four contestants attended six IMOs, 42 contestants attended five IMOs, and 221 contestants attended four IMOs, as shown in Table 13.

Table 13 Contestants with Six or More Participations in the First 64 IMOs

Contestant	Country	Partici-pation Year	Gold	Silver	Bronze	Honorable Mention	Perfect Score
David Kunszenti-Kovács	Norway	1997–2003	1	3	1	1	
Yeoh Zi Song	Malaysia	2014–2020	1	1	4	1	
Zhuo Qun (Alex) Song	Canada	2010–2015	5	0	1	0	1
Teodor von Burg	Serbia	2007–2012	4	1	1	0	
Alexey Entin	Israel	2000–2005	1	3	1	0	
Tan Li Xuan	Malaysia	2016–2021	0	2	2	1	

Coincidentally, in the 43rd IMO held in 2002, the gold medal cutoff was set at 29 points, and David Kunszenti-Kovács achieved a total score of exactly 29 points.

In the 44th IMO held in 2003, the gold medal cutoff was set at 29 points, whereas Alexey Entin attained a total score of exactly 28 points.

In the 51st IMO held in 2010, the bronze medal cutoff was set at 15 points, and Zhuo Qun (Alex) Song achieved a total score of exactly 15 points.

In the 58th IMO held in 2017, the silver medal cutoff was set at 19 points, and Yeoh Zi Song achieved a total score of exactly 19 points. In the 59th IMO held in 2018, the silver medal cutoff was set at 25 points, and Yeoh Zi Song's total score was exactly 24 points. In the 61st IMO held in 2020, the gold medal cutoff was set at 31 points, and Yeoh Zi Song's total score was exactly 31 points.

Additionally, from 2002 to 2005 and in 2007, Sherry Gong participated in the IMO, earning one bronze, two silver, and one gold medal. In the 48th IMO held in 2007, she ranked 7th individually. Notably, from 2002 to 2004, she was a member of the Chinese Puerto Rico IMO team, while in 2005 and 2007, she was a member of the United States IMO team.

From 2020 to 2023, Alex Chui participated in the IMO, securing two gold and two silver medals. However, in 2020 and 2021, he was a member of the Chinese Hong Kong IMO team, while in 2022 and 2023, he was a member of the United Kingdom IMO team.

Other than Sherry Gong and Alex Chui, the remaining 267 contestants hailed from 75 different countries and regions. Among them, there were

12 contestants from Cyprus and Moldova each, 11 from Malaysia, eight from Trinidad and Tobago, seven from each of Estonia, Germany, Sri Lanka, and North Macedonia, and six from Japan and Philippines each.

3.2 *Gold medals*

In the first 64 IMOs, a total of 49 contestants achieved three or more gold medals. Among them, one contestant earned five gold medals, six contestants earned four gold medals, and 42 contestants earned three gold medals.

As shown in Table 14, Zhuo Qun (Alex) Song, Reid Barton, and Lisa Sauermann have all achieved perfect scores.

Table 14   Contestants with Four or More Gold Medals in the First 64 IMOs

Contestant	Country	Participation Year	Gold Year	Perfect Score Year
Zhuo Qun (Alex) Song	Canada	2010–2015	2011–2015	2015
Reid Barton	The United States of America	1998–2001	1998–2001	2001
Christian Reiher	Germany	1999–2003	2000–2003	
Lisa Sauermann	Germany	2007–2011	2008–2011	2011
Teodor von Burg	Serbia	2007–2012	2009–2012	
Nipun Pitimanaaree	Thailand	2009–2013	2010–2013	
Luke Robitaille	The United States of America	2019–2022	2019–2022	

Coincidentally, in the 54th IMO held in 2013, the gold medal cutoff was 31 points, and Nipun Pitimanaaree achieved a total score of exactly 31 points.

In the 41st IMO held in 2000, the top four contestants all achieved perfect scores, and Reid Barton ranked fifth with a total score of 39. In the 43rd IMO held in 2002, the top three contestants all achieved perfect scores, and Christian Reiher ranked fourth with a total score of 36. Furthermore, in the 50th IMO held in 2009, Lisa Sauermann achieved a total score of 41, securing the third position.

Moreover, Oleg Golberg participated in the IMO from 2002 to 2004, achieving three gold medals. He consistently ranked within the top 10 in terms of total scores. Notably, in 2002 and 2003, he was a member of the Russia IMO team, and in 2004, he was a member of the United States IMO team.

Apart from Oleg Golberg, the remaining 48 contestants hailed from 22 different countries and regions. Among them, there were four contestants from each of Russia, Bulgaria, Germany, Hungary, Romania, and the United States. Both South Korea and the United Kingdom had three contestants, while Canada, Japan, Singapore, and the Soviet Union were represented by two contestants each.

3.3 *Special prizes*

In the first 64 IMOs, only 44 contestants received special prizes. Among them, one contestant has received the special prize three times, seven contestants have earned twice, and 36 contestants have achieved once. It indicates that achieving a special prize is even more challenging than securing a gold medal.

As shown in Table 15, John Rickard, Imre Ruzsa, and Marc van Leeuwen all achieved two special prizes in one IMO for their elegant solutions. Furthermore, John Rickard, Imre Ruzsa, and László Lovász have all earned a perfect score twice.

Table 15 Contestants with Multiple Special Prizes in the First 64 IMOs

Contestant	Country	Partici-pation Year	Special Prize Year	Gold Year	Perfect Score Year
John Rickard	The United Kingdom	1975–1977	1976, 1977 (2)	1975–1977	1975, 1977
József Pelikán	Hungary	1963–1966	1965, 1966	1964–1966	1966
László Lovász	Hungary	1963–1966	1965, 1966	1964–1966	1965, 1966
László Babai	Hungary	1966–1968	1966, 1968	1968	1968
Simon Phillips Norton	The United Kingdom	1967–1969	1967, 1969	1967–1969	1969
Wolfgang Burmeister	The German Democratic Republic	1967–1971	1970, 1971	1968, 1970, 1971	1970
Imre Ruzsa	Hungary	1969–1971	1971 (2)	1970, 1971	1970, 1971
Marc van Leeuwen	The Netherlands	1977, 1978	1978 (2)		

Coincidentally, in the 11th IMO held in 1969, only three gold medals were awarded, with Imre Ruzsa ranking 4th and receiving a silver medal.

Similarly, in the 19th IMO held in 1977, which also resulted in only 13 gold medals, Marc van Leeuwen ranked 14th and earned a silver medal.

Additionally, these 44 contestants hailed from 16 different countries and regions. Among them, there were seven contestants from each of Hungary and the United Kingdom, five from the German Democratic Republic, four from Bulgaria and Poland each, three from each of Czechoslovakia and the Soviet Union, and two from Finland and the United States each.

Special prizes were more frequently granted in the first 20 IMOs, with a total of 27 special prizes earned by 22 contestants from the 11th–20th IMOs. Subsequently, the frequency of special prize presentations declined. Since Moldovan contestant Iurie Boreico received a special prize for his brilliant solution to IMO 46-3 (Algebra, proposed by South Korea) in 2005, no contestant has achieved this accolade to date.

From 2003 to 2007, Iurie Boreico consistently participated in the IMO, earning three gold and two silver medals. He achieved a perfect score in 2005 and 2006. It's noteworthy that the 44th IMO held in 2003 only yielded 37 gold medals, with Iurie Boreico placing the 38th individually and receiving a silver medal.

- **(IMO 46-3, proposed by South Korea).** Let x, y, z be three positive reals such that $xyz \geq 1$. Prove that
$$\frac{x^5 - x^2}{x^5 + y^2 + z^2} + \frac{y^5 - y^2}{y^5 + z^2 + x^2} + \frac{z^5 - z^2}{z^5 + x^2 + y^2} \geq 0.$$

4 Summary

The IMO stands as a distinguished intellectual competition for young minds. According to a study by Agarwal R. and Gaule P., statistical analysis reveals that among contestants in the IMO (including those who did not secure medals), 22% choose to pursue further studies in mathematics, ultimately obtaining doctoral degrees in the field. Additionally, 1% of these contestants become presenters at the International Congress of Mathematicians, and 0.2% attain the Fields Medal. These statistics underscore the vital role of the IMO in identifying and nurturing mathematical talent.

It's essential not to perceive the IMO as a mere selection exam. Rather than focusing solely on the brief two-day competition, the crucial aspect lies in the learning and preparation undertaken before participating. As the mathematician Paul Halmos aptly put it, what mathematics really consists of is problems and solutions. Contestants, through their exploration of Olympiad problems, not only enhance their mathematical abilities but also experience the joy and satisfaction of problem-solving. This experience plants the seeds of a future career in mathematics.

However, it's important to acknowledge that Olympiad problems and research problems in mathematics differ. Research problems often lack readily available answers and may require the investment of countless days and nights. Hence, the IMO is just one pathway in the growth of mathematical talents, and success in the IMO is not the sole qualification for becoming an outstanding mathematician.

Although every contestant aims for a gold medal, their aspirations go far beyond accolades. On this stage, they have the opportunity to showcase their intellectual capabilities, revel in the mathematical exploration, and relish competing with talented young minds from around the world, all without the narrow goal of proving their superiority over others. While the competition results may vary, each contestant stands as a victor in their own right and becomes a companion and witness to one another's life journeys.

In contrast to the Olympics, where athletes' careers are closely intertwined with the Games, the IMO is merely a chapter in the growth of these gifted young individuals. Following the IMO, the door to a new mathematical world has already swung wide open for them.

Chapter 1

Enumerative Combinatorics Problems

Enumerative combinatorics problems are closely related to real-life applications and permeate various stages of mathematical education. From the principles of addition and multiplication in elementary school mathematics to permutations and combinations in high school mathematics, many enumerative combinatorics problems arise. It is not an exaggeration to say that enumerative combinatorics is the foundation of combinatorics mathematics and indeed of all mathematics domains.

In the first 64 IMOs, there had been a total of 19 enumerative combinatorics problems, accounting for approximately 21.8% of all combinatorics problems. The statistical distribution of these problems in the previous IMOs is presented in Table 1.1.

As seen in Table 1.1, enumerative combinatorics problems frequently appear in IMOs. In early IMOs, these problems were primarily simple calculation tasks with lower difficulty levels. In recent years, however, enumerative combinatorics problems have often been linked with set problems, requiring applications of various combinatorics methods based on properties of sets, which increases the difficulty.

Common techniques used in solving enumerative combinatorics problems include recursive methods, set correspondence, the inclusion–exclusion principle, and generating functions.

It is important to note that for each problem, the solutions are followed by information on the scores, including the number of contestants in

Table 1.1 Numbers of Enumerative Combinatorics Problems in the First 64 IMOs

Content	Session							Total
	1–10	11–20	21–30	31–40	41–50	51–60	61–64	
Enumerative combinatorics problems	2	1	7	2	4	3	0	19
Combinatorics problems	7	11	16	16	11	19	7	87
Percentage of enumerative combinatorics problems among combinatorics problems	28.6%	9.1%	43.8%	12.5%	36.4%	15.8%	0.0%	21.8%

each score range, the average score, and the scores of the top five teams. However, early IMOs often lacked information on contestant scores, so the number of contestants in each score range only represents the counted number of contestants, and some problems lack scores of the top five teams.

1.1 Common Theorems, Formulas, and Methods

1.1.1 *Preliminaries*

(1) *Principle of addition and multiplication*

(i) **Principle of Addition:** The methods for completing a task can be divided into n disjoint classes. In the first class, there are m_1 different methods, in the second class, there are m_2 different methods, ..., and in the nth class, there are m_n different methods. Then the total number of ways to complete the task is

$$m_1 + m_2 + \cdots + m_n.$$

(ii) **Principle of Multiplication:** If there are m_1 different ways to do the first thing, m_2 different ways to do the second thing after the first thing is done, ..., and m_n different ways to do the nth thing after the first, second, ..., and the $(n-1)$th things are done, then the number of ways to do the first thing, then the second thing, ..., and finally the nth thing is

$$m_1 \times m_2 \times \cdots \times m_n.$$

(2) *Arrangements and combinations without repetition*

(i) **Arrangements without Repetition:** We call a sequence of $k(1 \leq k \leq n)$ distinct elements from an n-element set a k arrangement of an n-element set. The number of k arrangements of an n-element set is denoted by P_n^k, and

$$P_n^k = n(n-1)\cdots(n-k+1) = \frac{n!}{(n-k)!}.$$

For $k = n$, we obtain the number of total arrangements:

$$P_n^n = n \cdot (n-1) \cdot \cdots \cdot 2 \cdot 1 = n!.$$

(ii) **Combinations without Repetition:** We call a $k(1 \leq k \leq n)$ elements subset of an n-element set a k combination of an n-element set. The number of k combinations of an n-element set is denoted by C_n^k, and

$$C_n^k = \frac{P_n^k}{k!} = \frac{n(n-1)\cdots(n-k+1)}{k!} = \frac{n!}{k!(n-k)!}.$$

(3) *Arrangements and combinations with repetition*

(i) **Arrangements with Repetition:** A sequence of $k(k \geq 1)$ elements (possibly equal) from an n-element set is called a k-element arrangement with repetition from an n-element set. The number of such arrangements is n^k.

(ii) **Partition into k Parts:** We partition an n-element set into k groups, with $n_1, n_2, \ldots, n_k$ elements respectively. The number of such partitions is

$$\frac{n!}{n_1! n_2! \cdots n_k!}.$$

(iii) **Combinations with Repetition:** A group of k elements (possibly equal) of an n-element set is called a $k(k \geq 1)$ combination with repetition of an n-element set. The number of such combinations is C_{n+k-1}^k.

(4) *Arrangements on a circle*

Taking k elements from n different elements (without repetition) to arrange them on a circle is called a k-circle arrangement of n different elements. If one arrangement can be obtained from another arrangement by rotation, they are considered the same circle arrangement. The number of different k-circle arrangements of n different elements is

$$\frac{\mathrm{P}_n^k}{k} = \frac{n!}{k \cdot (n-k)!}.$$

(5) *Inclusion–exclusion principle*

Let $S_1, \ldots, S_n$ be n finite sets. Then

$$\left| S_1 \bigcup \cdots \bigcup S_n \right| = \sum_{1 \le i \le n} |S_i| - \sum_{1 \le i_1 < i_2 \le n} \left| S_{i_1} \bigcap S_{i_2} \right| + \cdots$$

$$+ (-1)^{k-1} \sum_{1 \le i_1 < \cdots < i_k \le n} \left| S_{i_1} \bigcap \cdots \bigcap S_{i_k} \right| + \cdots$$

$$+ (-1)^{n-1} \left| S_1 \bigcap \cdots \bigcap S_n \right|.$$

Example 1.1 (Disarrangements). A total arrangement $(i_1, i_2, \ldots, i_n)$ of $\{1, 2, \ldots, n\}$ is called a disarrangement if $i_1 \ne 1, i_2 \ne 2, \ldots, i_n \ne n$. Find the number D_n of disarrangements of $\{1, 2, \ldots, n\}$.

Solution. Let S be the set of all permutations of $\{1, 2, \ldots, n\}$. Clearly $|S| = n!$. Let

$$A_j = \{(i_1, i_2, \ldots, i_n) \mid (i_1, i_2, \ldots, i_n) \in S, i_j = j\} \, (j = 1, 2, \ldots, n).$$

Thus $D_n = \left| \bar{A}_1 \bigcap \bar{A}_2 \bigcap \cdots \bigcap \bar{A}_n \right|$. It is easy to see that

$$|A_j| = (n-1)! (1 \le j \le n), \left| A_i \bigcap A_j \right| = (n-2)! (1 \le i < j \le n), \ldots,$$

$$\left| A_{i_1} \bigcap A_{i_2} \bigcap \cdots \bigcap A_{i_r} \right| = (n-r)! \, (1 \le i_1 < i_2 < \cdots < i_r \le n), \ldots,$$

$$\left| A_1 \bigcap A_2 \bigcap \cdots \bigcap A_n \right| = 0! = 1.$$

By the inclusion–exclusion principle,

$$D_n = |S| - \sum_{i=1}^{n} |A_i| + \sum_{1 \le i < j \le n} \left| A_i \bigcap A_j \right| - \cdots + (-1)^n$$

$$\times \left| A_1 \bigcap A_2 \bigcap \cdots \bigcap A_n \right|$$

$$= n! - \mathrm{C}_n^1 (n-1)! + \mathrm{C}_n^2 (n-2)! - \mathrm{C}_n^3 (n-3)! + \cdots + (-1)^n \mathrm{C}_n^n \cdot 0!$$

$$= n! \left(1 - \frac{1}{1!} + \frac{1}{2!} - \frac{1}{3!} + \cdots + \frac{(-1)^n}{n!} \right).$$

Note. This problem is often referred to as Bernoulli's misplacement problem.

(6) *Double counting principle (Fubini's principle)*

The general double counting principle is to represent the same quantity in two different ways and establish an equality. The usual form to apply this principle is to count some tuples, which may be counted by fixing the first component, or by fixing the second component.

For example, suppose we have an $m \times n$ matrix

$$
\begin{pmatrix}
a_{11} & a_{12} & \cdots & a_{1n} \\
a_{21} & a_{22} & \cdots & a_{2n} \\
\vdots & \vdots & & \vdots \\
a_{m1} & a_{m2} & \cdots & a_{mn}
\end{pmatrix}.
$$

If we first calculate the sum of numbers in the ith row $r_i = \sum_{j=1}^{n} a_{ij}$ $(i = 1, 2, \ldots, m)$, then adding up these row sums, we get the total sum of the numbers in the matrix, $\sum_{i=1}^{m} r_i = \sum_{i=1}^{m} (\sum_{j=1}^{n} a_{ij})$.

Another way is to calculate the column sums first; denote the sum of the numbers in the jth column by $l_j = \sum_{i=1}^{m} a_{ij} (j = 1, 2, \ldots, n)$. Adding up these column sums, we also get the total sum of the numbers in the matrix,

$$
\sum_{j=1}^{n} l_j = \sum_{j=1}^{n} \left(\sum_{i=1}^{m} a_{ij} \right).
$$

Hence, $\sum_{i=1}^{m} r_i = \sum_{j=1}^{n} l_j$, that is

$$
\sum_{i=1}^{m} \left(\sum_{j=1}^{n} a_{ij} \right) = \sum_{j=1}^{n} \left(\sum_{i=1}^{m} a_{ij} \right).
$$

(7) *Generating functions*

The concept of generating functions is based on the idea of mapping a sequence of numbers to a function (which can be a polynomial or a formal power series), and by studying properties of this function, we obtain properties of the original sequence.

The generating function serves as a "bridge" that connects sequences with functions, and this idea is worth exploring in detail.

Let $a_0, a_1, a_2, \ldots$ be a sequence of numbers. We call the power series,

$$
g(x) = \sum_{k=0}^{+\infty} a_k x^k,
$$

the generating function of $\{a_n\}$.

Example 1.2. Let $a_0 = -1$, $a_1 = 1$, and $a_n = 2a_{n-1} + 3a_{n-2} + 3^n (n \geq 2)$. Find a_n.

Solution. Let $f(x) = a_0 + a_1 x + a_2 x^2 + \cdots + a_n x^n + \cdots$. Then
$$-2xf(x) = -2a_0 x - 2a_1 x^2 - \cdots - 2a_{n-1} x^n + \cdots$$
and $-3x^2 f(x) = -3a_0 x^2 - \cdots - 3a_{n-2} x^n + \cdots$.

Summing up the above three equalities and utilizing the assumption $a_0 = -1$, $a_1 = 1$, and $a_n = 2a_{n-1} + 3a_{n-2} + 3^n$, we get
$$(1 - 2x - 3x^2) f(x) = -1 + 3x + 3^2 x^2 + \cdots + 3^n x^n + \cdots$$
$$= -1 + \frac{3x}{1 - 3x} = \frac{6x - 1}{1 - 3x}.$$

Thus,
$$f(x) = \frac{6x - 1}{(1+x)(1-3x)^2} = \frac{A}{1+x} + \frac{B}{(1-3x)^2} + \frac{C}{1-3x}. \quad (*)$$

Multiplying both sides of (*) by $1 + x$ and setting $x = -1$, we get
$$A = \frac{6x - 1}{(1 - 3x)^2} \Big|_{x=-1} = -\frac{7}{16}.$$

Multiplying both sides of (*) by $(1 - 3x)^2$ and setting $x = \frac{1}{3}$, we obtain
$$B = \frac{6x - 1}{1 + x} \Big|_{x=\frac{1}{3}} = \frac{3}{4}.$$

Multiplying both sides of (*) by x and taking the limit $x \to \infty$, we see that
$$0 = \lim_{x \to \infty} \frac{x(6x - 1)}{(1+x)(1-3x)^2}$$
$$= \lim_{x \to \infty} \left(\frac{Ax}{1+x} + \frac{Bx}{(1-3x)^2} + \frac{Cx}{1-3x} \right)$$
$$= A - \frac{1}{3}C.$$

Hence, $C = 3A = -\frac{21}{16}$. So,
$$f(x) = -\frac{7}{16(1+x)} + \frac{3}{4(1-3x)^2} - \frac{21}{16(1-3x)}$$
$$= -\frac{7}{16} \sum_{n=0}^{\infty} (-1)^n x^n + \frac{3}{4} \sum_{n=0}^{\infty} C_{n+1}^1 3^n x^n - \frac{21}{16} \sum_{n=0}^{\infty} 3^n x^n$$
$$= \sum_{n=0}^{\infty} \left[\frac{(4n - 3) \cdot 3^{n+1} - 7(-1)^n}{16} \right] x^n.$$

and

$$a_n = \frac{1}{16}\left[(4n-3)\cdot 3^{n+1} - 7(-1)^n\right].$$

(8) *Miscellany*

Seating Problem (Lucas). There are $n(n \geq 3)$ couples sitting around a table, alternating between men of women. The number of ways of such a circle arrangement that each couple is not adjacent is

$$\sum_{r=0}^{n}(-1)^r \frac{2n}{2n-r}C_{2n-r}^r(n-r)!$$

$$= n! - \frac{2n}{2n-1}C_{2n-1}^1(n-1)! + \frac{2n}{2n-2}C_{2n-2}^2(n-2)! - \cdots + 2(-1)^n.$$

Catalan Numbers I. There are $2n$ people waiting in a line before a sales booth for shopping, assuming they all purchase the same item which is worth 5 yuan. Among them, n people hold a 5 yuan banknote, while n people hold a 10 yuan banknote. At the beginning, the salesman had no change. The number of ways to queue the $2n$ people in a line such that the salesman can smoothly serve the shoppers without finding himself out of change is $\frac{1}{n+1}C_{2n}^n$. (*Note:* The n people with a 5 yuan banknote are not distinguishable, the same for the n people with a 10 yuan banknote.)

Catalan Numbers II (Dunkel). The number of ways to divide a convex n-gon into $n-2$ triangles by non-crossing diagonals is $\frac{1}{n-1}C_{2n-4}^{n-2}$.

Catalan Numbers III. There are $2n$ points evenly distributed on a circumference, and the number of ways to pair up n non-intersecting strings is $\frac{1}{n+1}C_{2n}^n$.

1.1.2 *Combinatorial identities*

(1) *Binomial theorem*

We have $(x+y)^n = \sum_{k=0}^{n} C_n^k y^k x^{n-k}$, and in particular, $(1+x)^n = \sum_{k=0}^{n} C_n^k x^k$, where n is a positive integer and $C_n^k = \frac{n(n-1)\cdots(n-k+1)}{k!}$ $(1 \leq k \leq n)$, with $C_n^0 = 1$.

(2) *Basic combinatorial identities*

    (i) $C_m^n = C_m^{m-n}$.
    (ii) $C_{m+1}^n = C_m^n + C_m^{n-1}$.

(iii) $C_n^k = \frac{n}{k} C_{n-1}^{k-1}$.

(iv) $C_n^k C_k^m = C_n^m C_{n-m}^{k-m}$.

 (v) $1 + C_n^1 + \cdots + C_n^n = 2^n$.

(vi) $1 - C_n^1 + C_n^2 - \cdots + (-1)^n C_n^n = 0$.

(vii) $C_n^1 + C_n^3 + \cdots + C_n^{2\left[\frac{n}{2}\right]+1} = 1 + C_n^2 + \cdots + C_n^{2\left[\frac{n}{2}\right]} = 2^{n-1}$.

(viii) $C_n^0 + C_{n+1}^1 + C_{n+2}^2 \cdots + C_{n+k}^k = C_{n+k+1}^k$.

(ix) **(Vandermonde's Identity).**

$$C_m^0 C_n^k + C_m^1 C_n^{k-1} + C_m^2 C_n^{k-2} + \cdots + C_m^k C_n^0 = C_{m+n}^k.$$

Example 1.3. Prove that $\sum_{k=1}^n k C_n^k = n \cdot 2^{n-1}$.

Proof. Since $k C_n^k = n C_{n-1}^{k-1}$, we have

$$\sum_{k=1}^n k C_n^k = \sum_{k=1}^n n C_{n-1}^{k-1} = n \sum_{k=0}^{n-1} C_{n-1}^k = n \cdot 2^{n-1}.$$

Example 1.4. Prove that $\sum_{k=0}^n C_{2n}^k = 2^{2n-1} + \frac{(2n)!}{2 \cdot (n!)^2}$.

Proof. We have

$$\sum_{k=0}^n C_{2n}^k = \sum_{k=0}^{2n} C_{2n}^k - \sum_{k=n+1}^{2n} C_{2n}^k = 2^{2n} - \sum_{k=n+1}^{2n} C_{2n}^k.$$

Substituting $k = 2n - j$ in the summation $\sum_{k=n+1}^{2n} C_{2n}^k$, we get

$$\sum_{k=n+1}^{2n} C_{2n}^k = \sum_{j=0}^{n-1} C_{2n}^{2n-j} = \sum_{j=0}^{n-1} C_{2n}^j = \sum_{j=0}^n C_{2n}^j - C_{2n}^n.$$

Thus,

$$\sum_{k=0}^n C_{2n}^k = 2^{2n} - \left(\sum_{k=0}^n C_{2n}^k - C_{2n}^n \right).$$

It follows that

$$\sum_{k=0}^n C_{2n}^k = \frac{1}{2} \left(2^{2n} + C_{2n}^n \right) = 2^{2n-1} + \frac{(2n)!}{2(n!)^2}.$$

Example 1.5. Prove that $C_r^r + C_{r+1}^r + \cdots + C_{r+n-1}^r = C_{r+n}^{r+1}$.

Proof. By Pascal's identity,

$$C_r^r + C_{r+1}^r + \cdots + C_{r+n-1}^r = C_{r+1}^{r+1} + \left(C_{r+2}^{r+1} - C_{r+1}^{r+1} \right) + \left(C_{r+3}^{r+1} - C_{r+2}^{r+1} \right)$$

$$+ \cdots + \left(C_{r+n}^{r+1} - C_{r+n-1}^{r+1} \right) = C_{r+n}^{r+1}.$$

(3) *Miscellany*

Theorem 1.1 (Multinomial Theorem). The coefficient of the monomial $x_1^{r_1} x_2^{r_2} \ldots x_k^{r_k}$ in the polynomial $(x_1 + x_2 + \cdots + x_k)^n$ is $\dfrac{n!}{r_1! r_2! \cdots r_k!}$, where $n = r_1 + r_2 + \cdots + r_k$.

This is a generalization of the binomial theorem.

After simplification, there are C_{n+k-1}^n monomials in $(x_1 + x_2 + \cdots + x_k)^n$.

Theorem 1.2 (Li Shanlan's Identity). $\displaystyle\sum_{j\geq 0} C_k^j C_l^j C_{n+k+l-j}^{k+l} = C_{n+k}^k C_{n+l}^l$.

1.1.3 *Common methods*

(1) *Enumeration*

The most elementary way for counting is to list the objects in detail and count.

Example 1.6. How many ways exist to represent 22 as a sum of two prime numbers? (Orders are not considered)

Solution. Assume $a + b = 22$, where a, b are primes and $a \leq b$. Therefore, $a \leq 11$, so $a = 2, 3, 5, 7,$ or 11. By a straightforward verification, we have three solutions $22 = 3 + 19 = 5 + 17 = 11 + 11$.

(2) *Principle of addition and multiplication*

Example 1.7. How many six-digit numbers contain both digit 2 and digit 6 exactly once?

Solution. Note that the leading digit is not zero.

If the leading digit is not 2 nor 6, then there are 7 choices for the leading digit, and for each of other three digits which are not 2 nor 6, there are 8 choices. In this case, there are $(P_5^2 \times 8^3 \times 7 =)71680$ such numbers.

If the leading digit is 2 or 6, then there are $(2 \times C_5^1 \times 8^4 =)40960$ such numbers.

In conclusion, the number of prescribed 6-digit numbers is $(71680 + 40960 =)112640$.

(3) *Double counting*

Example 1.8. Alice dissects a cube of side length 4 into 29 small cubes with integer side lengths. How many unit cubes are there?

Solution. If there is a small cube of side length 3, then all the remaining small cubes are unit cubes, and there are $4^3 - 3^3 = 37$ such cubes, which is not what we want.

Assume there are x small cubes of side length 2 and y unit cubes. Then

$$\begin{cases} x + y = 29 \\ 2^3 \cdot x + y = 4^3 \end{cases} \Rightarrow \begin{cases} x = 5 \\ y = 24. \end{cases}$$

This is possible, since we can first dissect the cube into 8 small cubes of side length 2, and then dissect 3 of them into 24 unit cubes. The answer is 24.

(4) *Recursion*

Some counting problems can be interpreted as finding a sequence $a_1, a_2, \ldots, a_n, \ldots$. It is usually difficult to find the expression of a_n straight-forwardly. If we can find a recursive relation (e.g., a relation between a_n and a_{n-1}), we may find the solution by solving the recursive sequence.

Example 1.9. Find the maximum number of regions that a plane can be divided into by 100 straight lines.

Solution. Assume that the plane is divided into a maximum a_n regions by n lines. It is easy to see that $a_1 = 2$, $a_2 = 4 = a_1 + 2$, and $a_3 = 7 = a_2 + 3$. In general, $a_n = a_{n-1} + n$. (In fact, assume we have $n-1$ lines already which divide the plane into a_{n-1} regions. To get as many regions as possible, the nth line l_n should intersect each previous lines, and the intersections do not coincide. The line l_n has $n - 1$ intersections which divide l_n into n parts, each part dividing some region into two regions. Thus we have n more regions, and hence $a_n = a_{n-1} + n$.) So

$$a_{100} = a_{99} + 100 = a_{98} + 99 + 100 = a_{97} + 98 + 99 + 100$$

$$= a_1 + 2 + 3 + \cdots + 100$$

$$= 2 + 2 + 3 \cdots + 100 = 1 + (1 + 2 + \cdots + 100)$$

$$= 1 + \frac{(1 + 100) \times 100}{2} = 1 + 5050 = 5051.$$

(5) *Inclusion–exclusion principle*

Example 1.10. How many positive integers are coprime to 105 from 1 to 1000?

Solution. Since $105 = 3 \times 5 \times 7$, a positive integer is coprime to 105 if and only if it is not divisible by 3, 5, and 7. Let $I = \{1, 2, \ldots, 1000\}$, and let A_3, A_5, and A_7 be the sets of numbers in I which are divisible by 3, 5, and 7 respectively. And denote by $\bar{A}_3$, $\bar{A}_5$, and $\bar{A}_7$ the complement of A_3, A_5, and A_7 in I, respectively (i.e., $\bar{A}_3$, $\bar{A}_5$, and $\bar{A}_7$ are the sets of the numbers in I which are not divisible by 3, 5, and 7, respectively). Thus, the set of numbers in I coprime to 105 is exactly $\bar{A}_3 \cap \bar{A}_5 \cap \bar{A}_7$. By the inclusion–exclusion principle,

$$
\begin{aligned}
\left| \bar{A}_3 \cap \bar{A}_5 \cap \bar{A}_7 \right| &= |I| - \left| A_3 \bigcup A_5 \bigcup A_7 \right| \\
&= |I| - \left(|A_3| + |A_5| + |A_7| \right) + \left(\left| A_3 \cap A_5 \right| + \left| A_3 \cap A_7 \right| \right. \\
&\quad \left. + \left| A_5 \cap A_7 \right| \right) - \left| A_3 \cap A_5 \cap A_7 \right| \\
&= 1000 - \left(\left[\frac{1000}{3} \right] + \left[\frac{1000}{5} \right] + \left[\frac{1000}{7} \right] \right) \\
&\quad + \left(\left[\frac{1000}{3 \times 5} \right] + \left[\frac{1000}{3 \times 7} \right] + \left[\frac{1000}{5 \times 7} \right] \right) - \left[\frac{1000}{3 \times 5 \times 7} \right] \\
&= 1000 - (333 + 200 + 142) + (66 + 47 + 28) - 9 \\
&= 1000 - 675 + 141 - 9 = 457.
\end{aligned}
$$

That is, there are 457 numbers from 1 to 1000 which are coprime to 105.

(6) *Correspondence*

When it is difficult to calculate the number of elements in a set M, we try to establish a one-to-one correspondence between the set M and another set N. If it is easier to calculate the number of elements in N, we can get the number of elements in M since $|M| = |N|$. This is the basic idea of using the correspondence principle to solve counting problems.

Example 1.11. There are 20 players who participate in a table tennis tournament. For each match, one player is eliminated. How many matches do we need to play in total to create a champion?

Solution. Note that each match eliminates exactly one player. To have a winner, we need to eliminate 19 players, thus we need 19 matches in total.

1.2   Problems and Solutions

Problem 1.1 (IMO 8-1, proposed by the Soviet Union). In a mathematical contest, three problems, A, B, and C were posed. Among the participants there were 25 students who solved at least one problem each. Of all the contestants who did not solve problem A, the number of students who solved B was twice the number of those who solved C. The number of students who solved only problem A was one more than the number of students who solved A and at least one other problem. Of all students who solved just one problem, half did not solve problem A. How many students solved only problem B?

Solution. Let x_A, x_B, x_C, x_{AB}, x_{BC}, x_{CA}, and x_{ABC} denote the number of students who solved the problems as in the subscripts. It follows from the assumption that

$$x_A + x_B + x_C + x_{AB} + x_{BC} + x_{CA} + x_{ABC} = 25. \tag{1}$$

Among the $x_B + x_C + x_{BC}$ students who did not solve problem A, there are $x_B + x_{BC}$ students who solved problem B and $x_C + x_{BC}$ students who solved problem C. Hence,

$$x_B + x_{BC} = 2(x_C + x_{BC}),$$

i.e., $$x_B - 2x_C - x_{BC} = 0. \tag{2}$$

There are x_A students who solved only problem A and there are $x_{AB} + x_{AC} + x_{ABC}$ students who solved problem A and at least one other problem. Thus

$$x_A = x_{AB} + x_{AC} + x_{ABC} + 1. \tag{3}$$

Since among all students who solved just one problem, half did not solve problem A, we have

$$x_A = x_B + x_C. \tag{4}$$

From (1) + (3) + (4)

$$3x_A + x_{BC} = 26, \tag{5}$$

and thus, $x_A \leq 8$.

It follows from (2), (4), and (5) that

$$x_A + 3x_B = 26 \quad \text{and} \quad x_B = \frac{26 - x_A}{3}.$$

Thus, $x_A = 2, 5, 8$.

If $x_A = 2$, then $x_B = 8$, contrary to (4).

If $x_A = 5$, then $x_B = 7$, contrary to (4).

If $x_A = 8$, then $x_B = 6$, $x_{BC} = 2$, $x_C = 2$, and $x_{AB} + x_{AC} + x_{ABC} = 7$, which fulfills all requirements. Therefore, there are six students who solved only problem B.

【Score Situation】This particular problem saw the following distribution of scores among contestants: 22 contestants scored 6 points, no contestant scored 5 points, 2 contestants scored 4 points, 1 contestant scored 3 points, no contestant scored 2 points, 1 contestant scored 1 point, and 1 contestant scored 0 point. The average score of this problem is 5.333, indicating that it was simple.

Among the top five teams in the team scores, the Soviet Union team achieved a total score of 293 points, the Hungary team achieved a total score of 281 points, the German Democratic Republic team achieved a total score of 280 points, the Poland team achieved a total score of 269 points, and the Romania team achieved a total score of 257 points.

The gold medal cutoff for this IMO was set at 39 points (with 13 contestants earning gold medals), the silver medal cutoff was 34 points (with 15 contestants earning silver medals), and the bronze medal cutoff was 31 points (with 11 contestants earning bronze medals).

In this IMO, a total of 11 contestants achieved a perfect score of 40 points.

Problem 1.2 (IMO 9-6, proposed by Hungary). In a sports contest, there were m medals awarded on n successive days ($n > 1$). On the first day, one medal and $\frac{1}{7}$ of the remaining $m - 1$ medals were awarded. On the second day, two medals and $\frac{1}{7}$ of the now remaining medals were awarded; and so on. On the nth and last day, the remaining n medals were awarded. How many days did the contest last, and how many medals were awarded altogether?

Solution. Let u_k be the number of medals remaining before the kth day. On the kth day, a total of $k + \frac{1}{7}(u_k - k)$ medals were awarded, and

$$u_k - \left[k + \frac{1}{7}(u_k - k) \right] = \frac{6}{7}(u_k - k)$$

medals were left. Therefore we have the following recursive formula

$$u_{k+1} = \frac{6}{7}(u_k - k).$$

Since $u_n = n$ and $u_1 = m$,

$$m - \frac{7}{6}u_2 = 1, u_2 - \frac{7}{6}u_3 = 2, u_3 - \frac{7}{6}u_4 = 3,$$

$$\cdots\cdots\cdots\cdots\cdots\cdots\cdots\cdots\cdots\cdots\cdots\cdots\cdots$$

$$u_{n-2} - \frac{7}{6}u_{n-1} = n - 2, u_{n-1} - \frac{7}{6}n = n - 1.$$

Multiplying the above equations by $1, \frac{7}{6}, \left(\frac{7}{6}\right)^2, \ldots, \left(\frac{7}{6}\right)^{n-2}$, respectively and adding up, we get

$$m = 1 + 2\left(\frac{7}{6}\right) + 3\left(\frac{7}{6}\right)^2 + \cdots + (n-1)\left(\frac{7}{6}\right)^{n-2} + n\left(\frac{7}{6}\right)^{n-1}. \quad (1)$$

Hence,

$$\frac{7}{6}m = \frac{7}{6} + 2\left(\frac{7}{6}\right)^2 + 3\left(\frac{7}{6}\right)^3 + \cdots + n\left(\frac{7}{6}\right)^n. \quad (2)$$

Subtracting (2) from (1), we obtain

$$-\frac{1}{6}m = 1 + \frac{7}{6} + \left(\frac{7}{6}\right)^2 + \cdots + \left(\frac{7}{6}\right)^{n-1} - n\left(\frac{7}{6}\right)^n$$

$$= \frac{\left(\frac{7}{6}\right)^n - 1}{\frac{7}{6} - 1} - n\left(\frac{7}{6}\right)^n$$

$$= \left(\frac{7}{6}\right)^n (6 - n) - 6.$$

Thus,

$$m = \frac{7^n(n-6)}{6^{n-1}} + 36.$$

Since $n > 1$, we see that $|n - 6| < 6^{n-1}$, and 7^n is coprime to 6^{n-1}. Thus, we must have $n - 6 = 0$. This gives $n = 6$ and $m = 36$.

Remark. In the solution we actually obtained a method to solve a certain recursive sequence.

【Score Situation】 This particular problem saw the following distribution of scores among contestants: 14 contestants scored 8 points, 1 contestant scored 7 points, 5 contestants scored 6 points, 3 contestants scored 5 points, 1 contestant scored 4 points, 4 contestants scored 3 points, 3 contestants scored 2 points, 1 contestant scored 1 point, and 5 contestants scored 0 point. The average score of this problem is 5.054, indicating that it was simple.

Among the top five teams in the team scores, the Soviet Union team achieved a total score of 275 points, the German Democratic Republic team achieved a total score of 257 points, the Hungary team achieved a total score of 251 points, the United Kingdom team achieved a total score of 231 points, and the Romania team achieved a total score of 214 points.

The gold medal cutoff for this IMO was set at 38 points (with 11 contestants earning gold medals), the silver medal cutoff was 30 points (with 14 contestants earning silver medals), and the bronze medal cutoff was 22 points (with 26 contestants earning bronze medals).

In this IMO, a total of five contestants achieved a perfect score of 42 points.

Problem 1.3 (IMO 13-6, proposed by Sweden). Let $A = (a_{ij})(i, j = 1, 2, \ldots, n)$ be a square matrix whose entries are non-negative integers. Suppose that whenever an entry $a_{ij} = 0$, the sum of the entries in the ith row and jth column is at least n. Prove that the sum of all the entries of the matrix is at least $\frac{1}{2}n^2$.

Proof. Denote

$$R_i = \sum_{j=1}^{n} a_{ij}, \quad i = 1, 2, \ldots, n$$

and

$$C_j = \sum_{i=1}^{n} a_{ij}, \quad j = 1, 2, \ldots, n,$$

the sum of the ith row and jth column. Let

$$p = \min\{R_1, R_2, \ldots, R_n, C_1, C_2, \ldots, C_n\}.$$

Case 1: $p \geq \frac{n}{2}$. Then the sum of all entries of the matrix, denoted by S, is at least np. Hence,

$$S \geq np \geq \frac{1}{2}n^2.$$

Case 2: $p < \frac{n}{2}$. By rearranging rows and columns, we may assume that the sum of the first row is p, and non-zero entries of the first row are the first q entries. Since the q non-zero entries are positive integers, they add up to at least q, so we obtain

$$q \leq p < \frac{n}{2}.$$

Consider the $(q+1)$th column. Since its first entry is 0, it follows from the assumption that $C_{q+1} \geq n - p$. Similarly, the sum of all entries of each $(q+2)$th, $\ldots$, nth column is at lease $n - p$.

Thus, adding the sum of each column, we get

$$S \geq pq + (n-p)(n-q)$$

$$= \frac{n^2}{2} + \frac{1}{2}(n-2p)(n-2q)$$

$$\geq \frac{n^2}{2}.$$

【Score Situation】 This particular problem saw the following distribution of scores among contestants: 1 contestant scored 8 points, no contestant scored 7 points, no contestant scored 6 points, no contestant scored 5 points, no contestant scored 4 points, no contestant scored 3 points, 1 contestant scored 2 points, 4 contestants scored 1 point, and 22 contestants scored 0 point. The average score of this problem is 0.500, indicating that it was extremely difficult.

Among the top five teams in the team scores, the scores of this problem are as follows: the Hungary team scored 53 points (with a total team score of 255 points), the Soviet Union team scored 38 points (with a total team score of 205 points), the German Democratic Republic team scored 18 points (with a total team score of 142 points), the Poland team scored 8 points (with a total team score of 118 points) , the Romania team scored 10 points (with a total team score of 110 points), and the United Kingdom team scored 7 points (with a total team score of 110 points).

The gold medal cutoff for this IMO was set at 35 points (with 7 contestants earning gold medals), the silver medal cutoff was 23 points (with 12 contestants earning silver medals), and the bronze medal cutoff was 11 points (with 29 contestants earning bronze medals).

In this IMO, only one contestant achieved a perfect score of 42 points, namely Imre Ruzsa from Hungary.

Problem 1.4 (IMO 21-6, proposed by Germany). Let A and E be opposite vertices of a regular octagon. A frog starts jumping at vertex A. From any vertex of the octagon except for E, it may jump to either of the two adjacent vertices. When it reaches vertex E, the frog stops and stays there. Let e_n be the number of distinct paths of exactly n jumps ending at E. Prove that $e_{2n-1} = 0$, $e_{2n} = \frac{1}{\sqrt{2}}(x^{n-1} - y^{n-1})$, $n = 1, 2, \ldots$, where $x = 2 + \sqrt{2}$ and $y = 2 - \sqrt{2}$.

Note. A path of n jumps is a sequence of vertices $(P_0, P_1, \ldots, P_n)$ such that:

(i) $P_0 = A$, $P_n = E$;
(ii) for every i with $0 \leq i \leq n-1$, the vertex P_i is distinct from E;
(iii) for every i with $0 \leq i \leq n-1$, the vertices P_i and P_{i+1} are adjacent.

Proof. Denote the number of paths from A to B, C, D, and A by n jumps by b_n, c_n, d_n, and a_n, respectively. By symmetry, the number of paths from A to H, G, and F by n jumps is b_n, c_n, and d_n, respectively. Clearly

$$e_n = 2d_{n-1}, \tag{1}$$

$$c_n = b_{n-1} + d_{n-1}, \tag{2}$$

$$b_n = c_{n-1} + a_{n-1}, \tag{3}$$

$$a_n = 2b_{n-1}, \tag{4}$$

$$d_n = c_{n-1}. \tag{5}$$

It follows that

$$
\begin{aligned}
e_n &= 2d_{n-1} = 2c_{n-2} \\
&= 2\left(b_{n-3} + d_{n-3}\right) = 2\left(c_{n-4} + a_{n-4} + d_{n-3}\right) \\
&= 2\left(2d_{n-3} + 2b_{n-5}\right) = 2[2d_{n-3} + 2\left(d_{n-3} - d_{n-5}\right)] \\
&= 8d_{n-3} - 4d_{n-5} \\
&= 4e_{n-2} - 2e_{n-4}.
\end{aligned}
$$

The initial terms $e_1 = e_2 = e_3 = 0$ and $e_4 = 2$ satisfy the required formula.

We use induction to prove the statement. Assume that the statement is true for all $n \le k (k \ge 2)$. Then $e_{2k+1} = 4e_{2k-1} - 2e_{2k-3} = 0$ and

$$
\begin{aligned}
e_{2k+2} &= 4e_{2k} - 2e_{2k-2} \\
&= \frac{4}{\sqrt{2}}(x^{k-1} - y^{k-1}) - \frac{2}{\sqrt{2}}(x^{k-2} - y^{k-2}) \\
&= \frac{1}{\sqrt{2}}[(x+y)(x^{k-1} - y^{k-1}) - xy(x^{k-2} - y^{k-2})] \\
&= \frac{1}{\sqrt{2}}(x^k - y^k).
\end{aligned}
$$

Our proof is finished by induction.

【Score Situation】 This particular problem saw the following distribution of scores among contestants: 58 contestants scored 7 points, 4 contestants scored 6 points, 8 contestants scored 5 points, 8 contestants scored 4 points, 12 contestants scored 3 points, 17 contestants scored 2 points, 50 contestants scored 1 point, and 9 contestants scored 0 point. The average score of this problem is 3.747, indicating that it was relatively straightforward.

Among the top five teams in the team scores, the scores of this problem are as follows: the Soviet Union team scored 48 points (with a total team score of 267 points), the Romania team scored 49 points (with a total team score of 240 points), the Germany team scored 49 points (with a total team score of 235 points), the United Kingdom team scored 52 points (with a total team score of 218 points), and the United States team scored 52 points (with a total team score of 199 points).

The gold medal cutoff for this IMO was set at 37 points (with 8 contestants earning gold medals), the silver medal cutoff was 29 points (with 32 contestants earning silver medals), and the bronze medal cutoff was 20 points (with 42 contestants earning bronze medals).

In this IMO, a total of four contestants achieved a perfect score of 40 points.

Problem 1.5 (IMO 22-2, proposed by Germany). Let $1 \leq r \leq n$ and consider all subsets of r elements of the set $\{1, 2, \ldots, n\}$. Each of these subsets has a smallest member. Let $F(n, r)$ denote the arithmetic mean of these smallest numbers. Prove that $F(n, r) = \frac{n+1}{r+1}$.

Proof. The set $\{1, 2, \ldots, n\}$ has C_n^r r-subsets, among which the number of r-subsets with least element $k (1 \leq k \leq n - r + 1)$ is C_{n-k}^{r-1} (the remaining $r - 1$ elements are chosen from $k+1, k+2, \ldots, n$), so

$$F(n, r) = \frac{1}{C_n^r} \sum_{k=1}^{n-r+1} k C_{n-k}^{r-1}.$$

Since $C_{n-k}^r + C_{n-k}^{r-1} = C_{n-k+1}^r$,

$$k C_{n-k}^{r-1} = k C_{n-k+1}^r - k C_{n-k}^r = (k-1) C_{n-(k-1)}^r - k C_{n-k}^r + C_{n-(k-1)}^r.$$

Then

$$\sum_{k=1}^{n-r+1} k C_{n-k}^{r-1} = \sum_{k=1}^{n-r+1} [(k-1) C_{n-(k-1)}^r - k C_{n-k}^r] + \sum_{k=1}^{n-r+1} C_{n-(k-1)}^r$$

$$= \sum_{k=1}^{n-r+1} C_{n-k+1}^r$$

$$= \sum_{k=1}^{n-r+1} (C_{n-k+2}^{r+1} - C_{n-k+1}^{r+1})$$

$$= C_{n+1}^{r+1},$$

and we obtain $F(n, r) = \frac{1}{C_n^r} \cdot C_{n+1}^{r+1} = \frac{n+1}{r+1}$.

Remark. Another way to compute $\sum_{k=1}^{n-r+1} k C_{n-k}^{r-1}$ is as follows. The terms in the summation may be written as the following triangle:

$$
\begin{array}{cccc}
C_{n-1}^{r-1} & & & \\
C_{n-2}^{r-1} & C_{n-2}^{r-1} & & \\
C_{n-3}^{r-1} & C_{n-3}^{r-1} & C_{n-3}^{r-1} & \\
\vdots & \vdots & \vdots & \ddots \\
C_{r-1}^{r-1} & C_{r-1}^{r-1} & C_{r-1}^{r-1} & \cdots & C_{r-1}^{r-1}.
\end{array}
$$

The sum of each column is $C_n^r, C_{n-1}^r, \ldots, C_r^r$, and adding these numbers up, we get C_{n+1}^{r+1}.

【Score Situation】This particular problem saw the following distribution of scores among contestants: 42 contestants scored 7 points, 2 contestants scored 6 points, no contestant scored 5 points, 1 contestant scored 4 points, no contestant scored 3 points, 5 contestants scored 2 points, no contestant scored 1 point, and 1 contestant scored 0 point. The average score of this problem is 6.275, indicating that it was simple.

Among the top five teams in the team scores, the scores of this problem are as follows: the United States team scored 56 points (with a total team score of 314 points), the Germany team scored 54 points (with a total team score of 312 points), the United Kingdom team scored 54 points (with a total team score of 301 points), the Austria team scored 51 points (with a total team score of 290 points), and the Bulgaria team scored 47 points (with a total team score of 287 points).

The gold medal cutoff for this IMO was set at 41 points (with 36 contestants earning gold medals), the silver medal cutoff was 34 points (with 37 contestants earning silver medals), and the bronze medal cutoff was 26 points (with 30 contestants earning bronze medals).

In this IMO, a total of 26 contestants achieved a perfect score of 42 points.

Problem 1.6 (IMO 28-1, proposed by Germany). Let $P_n(k)$ be the number of permutations of the set $\{1, 2, \ldots, n\}$ with $n \geq 1$, which have exactly k fixed points. Prove that

$$
\sum_{k=0}^{n} k P_n(k) = n!
$$

(**Remark:** A permutation f of a set S is a one-to-one mapping of S onto itself. An element i in S is called a fixed point of the permutation f if $f(i) = i$.).

Proof 1. For computing $P_n(k)$, we have C_n^k choices for the k fixed points, and the permutation of the rest $n - k$ numbers has no fixed point and there are $P_{n-k}(0)$ such permutations. Thus

$$P_n(k) = P_{n-k}(0)C_n^k,$$

where $P_0(0) = 1$. Hence

$$\sum_{k=0}^{n} kP_n(k) = \sum_{k=1}^{n} kP_{n-k}(0)C_n^k$$

$$= \sum_{k=1}^{n} P_{n-k}(0) \cdot k \cdot \frac{n!}{k!(n-k)!}$$

$$= \sum_{k=1}^{n} P_{n-k}(0) \cdot \frac{n \cdot (n-1)!}{(k-1)!((n-1)-(k-1))!}$$

$$= n \sum_{k=1}^{n} P_{n-k}(0)C_{n-1}^{k-1}$$

$$= n \sum_{k=1}^{n} P_{n-1}(k-1).$$

The summation $\sum_{k=1}^{n} P_{n-1}(k-1)$ counts the permutations of $n-1$ numbers with no fixed points, 1 fixed point, $\ldots$, and $n-1$ fixed points, which counts exactly all permutations of $n-1$ numbers. Hence it is equal to $(n-1)!$. Consequently,

$$\sum_{k=0}^{n} kP_n(k) = n(n-1)! = n!.$$

Proof 2. We prove it by induction on n.

For $n = 1$, the statement is obvious. Assume that the statement is true for $n - 1$, that is

$$\sum_{k=0}^{n-1} kP_{n-1}(k) = (n-1)!.$$

Clearly, we also have

$$\sum_{k=0}^{n-1} P_{n-1}(k) = (n-1)!.$$

Consider the case of n: a permutation of $1, 2, \ldots, (n-1), n$ with k fixed points can be obtained from permutations of $1, 2, \ldots, (n-1)$ in the following ways.

(i) Choose one of $P_{n-1}(k-1)$ permutations of $1, 2, \ldots, (n-1)$ with $k-1$ fixed points. Put n in the end and obtain a permutation of $1, 2, \ldots, (n-1), n$ with k fixed points. The number of such permutations is $P_{n-1}(k-1)$.

(ii) Choose one of $P_{n-1}(k)$ permutations of $1, 2, \ldots, (n-1)$ with k fixed points. Replacing one of $n-k-1$ non-fixed points by n and moving that number to the nth position, we obtain a permutation of $1, 2, \ldots, (n-1), n$ with k fixed points. The number of such permutations is

$$C_{n-k-1}^1 P_{n-1}(k) = (n-k-1)P_{n-1}(k).$$

(iii) Choose one of $P_{n-1}(k+1)$ permutations of $1, 2, \ldots, (n-1)$ with $k+1(k \le n-2)$ fixed points. Replacing one of the fixed points by n and moving that number to the nth position, we obtain a permutation of $1, 2, \ldots, (n-1), n$ with k fixed points. The number of such permutations is

$$C_{k+1}^1 P_{n-1}(k+1) = (k+1)P_{n-1}(k+1).$$

From the above three cases, we obtain that

$$P_n(k) = P_{n-1}(k-1) + (n-(k+1))P_{n-1}(k) + (k+1)P_{n-1}(k+1). \quad (*)$$

Substituting $k = 0, 1, \ldots, n$ into $(*)$ and multiplying by k give

$$0P_n(0) = 0(n-1)P_{n-1}(0) + 0 \cdot 1P_{n-1}(1),$$

$$1P_n(1) = 1P_{n-1}(0) + 1(n-2)P_{n-1}(1) + 1 \cdot 2P_{n-1}(2),$$

$$2P_n(2) = 2P_{n-1}(1) + 2(n-3)P_{n-1}(2) + 2 \cdot 3P_{n-1}(3),$$

$$3P_n(3) = 3P_{n-1}(2) + 3(n-4)P_{n-1}(3) + 3 \cdot 4P_{n-1}(4),$$

$$\cdots\cdots\cdots\cdots\cdots\cdots\cdots\cdots\cdots\cdots\cdots\cdots\cdots\cdots\cdots$$

$$(n-2)P_n(n-2) = (n-2)P_{n-1}(n-3) + (n-2)(n-(n-1)).$$
$$P_{n-1}(n-2) + (n-2)(n-1)P_{n-1}(n-1),$$

$$(n-1)P_n(n-1) = (n-1)P_{n-1}(n-2),$$

$$nP_n(n) = nP_{n-1}(n-1).$$

Adding up the above equalities, we get

$$\sum_{k=0}^{n} k P_n(k) = \sum_{k=0}^{n-1}(k+1)P_{n-1}(k) + \sum_{k=0}^{n-1} k(n-k-1)P_{n-1}(k)$$

$$+ \sum_{k=1}^{n-1}(k-1)k P_{n-1}(k)$$

$$= \sum_{k=0}^{n-1}((k+1) + k(n-k-1) + (k-1)k)P_{n-1}(k)$$

$$= \sum_{k=0}^{n-1}(kn - k + 1)P_{n-1}(k)$$

$$= \sum_{k=0}^{n-1}(n-1)k P_{n-1}(k) + \sum_{k=0}^{n-1} P_{n-1}(k)$$

$$= (n-1)(n-1)! + (n-1)! = n!.$$

Proof 3. Let S_n be the set consisting of all permutations of $1, 2, \ldots, n$. We use $\pi = (\pi(1), \ldots, \pi(n))$ to denote one permutation. Consider the set

$$M = \{(\pi, i) \,|\, \pi \in S_n \,, 1 \le i \le n, \pi(i) = i\}.$$

Count the number of elements of M. For $\pi \in S_n$, denote the number of fixed points of π by $f(\pi)$. On the one side,

$$|M| = \sum_{\pi \in S_n} \sum_{(\pi,i) \in M} 1 = \sum_{\pi \in S_n} f(\pi) = \sum_{k=0}^{n} k P_n(k).$$

On the other side,

$$|M| = \sum_{i=1}^{n} \sum_{(\pi,i) \in M} 1 = \sum_{i=1}^{n}(n-1)! = n!.$$

(The number of permutations π with i being a fixed point is $(n-1)!$.) Thus,

$$\sum_{k=0}^{n} k P_n(k) = n!.$$

【Score Situation】 This particular problem saw the following distribution of scores among contestants: 98 contestants scored 7 points, 1 contestant scored 6 points, 7 contestants scored 5 points, 3 contestants scored 4 points, 7 contestants scored 3 points, 15 contestants scored 2 points, 31 contestants scored 1 point, and 75 contestants scored 0 point. The average score of this problem is 3.464, indicating that it was relatively straightforward.

Among the top five teams in the team scores, the scores of this problem are as follows: the Romania team scored 42 points (with a total team score of 250 points), the Germany team scored 42 points (with a total team score of 248 points), the Soviet Union team scored 42 points (with a total team score of 235 points), the German Democratic Republic team scored 42 points (with a total team score of 231 points), and the United States team scored 42 points (with a total team score of 220 points).

The gold medal cutoff for this IMO was set at 42 points (with 22 contestants earning gold medals), the silver medal cutoff was 32 points (with 42 contestants earning silver medals), and the bronze medal cutoff was 18 points (with 56 contestants earning bronze medals).

In this IMO, a total of 22 contestants achieved a perfect score of 42 points.

Problem 1.7 (IMO 29-2, proposed by Czechoslovakia). Let n be a positive integer and let $A_1, A_2, \ldots, A_{2n+1}$ be subsets of a set B. Suppose that:

(a) each A_i has exactly $2n$ elements;
(b) each $A_i \cap A_j (1 \leq i < j \leq 2n+1)$ contains exactly one element;
(c) every element of B belongs to at least two of the A_i.

For which values of n can one assign to every element of B one of the numbers 0 and 1 in such a way that A_i has 0 assigned to exactly n of its elements?

Solution 1. There exists such an assignment if and only if n is even.

First, we show that condition (c) can be strengthened as follows: every element of B belongs to exactly two sets $A_i (1 \leq i \leq 2n+1)$.

In fact, if there is some a of B belonging to 3 or more sets, say $a \in A_1 \cap A_2 \cap A_3$, by condition (b), each of the remaining $2n - 2$ sets $A_4, \ldots, A_{2n+1}$ has exactly one element of A_1. Therefore A_1 has an element not belonging to $A_2 \cup A_3 \cup \cdots \cup A_{2n+1}$, contradicting condition (c).

Let $V = \{A_1, A_2, \ldots, A_n\}$. Every element a of B belongs to exactly two sets A_i, A_j, joining an edge between A_i, A_j for a. In this way, the elements of B are in one to one correspondence with the edges of the complete graph K_V, and the elements of set A_i are exactly the $2n$ edges at A_i.

If there exists an assignment of numbers 0 and 1 to each element of B (i.e., the edges of K_V) such that every vertex A_i has half of the edges assigning 0, the number of edges assigning 0 is $\frac{1}{2}n(2n + 1)$, which is an integer, so n is even.

If n is even, then we construct such an assignment. Mark $2n+1$ points equally distributed on a circle with length $2n+1$, and label the points with $1, 2, \ldots, 2n+1$ in order (corresponding to $A_1, A_2, \ldots, A_{2n+1}$), as shown in Figure 1.1. For every pair of elements i, j of $\{1, 2, \ldots, 2n+1\}$, if the minor arc between them has an odd length, assign 1 to the edge between i and j, and assign 0 otherwise. Since n is even, exactly half of edges at A_i are assigned 0.

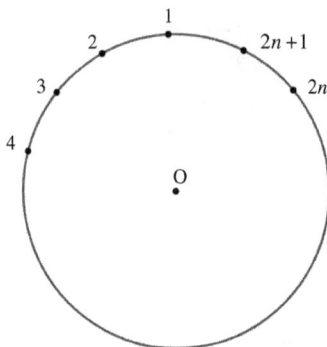

Figure 1.1   A Circle Equally Distributed

Solution 2. As in the first solution, we conclude that every element of A_i belongs to exactly one of the other $2n$ sets, and $B = A_1 \bigcup A_2 \bigcup \cdots \bigcup A_{2n+1}$. Construct a $2n \times 2n$ symmetric matrix (a_{ij}) with entries in B: for $1 \leq i < j \leq 2n$, $a_{ij} = a_{ji}$ is an element of $A_i \bigcap A_j$; for $1 \leq i \leq 2n$, a_{ii} is an element of $A_i \bigcap A_{2n+1}$. Thus, entries in the ith row (or column) of the matrix (a_{ij}) are elements of A_i and entries on the main diagonal are elements of A_{2n+1}.

If there exists the required assignment, then there are $2n \cdot n = 2n^2$ entries of (a_{ij}) assigned 0. Since (a_{ij}) is symmetric, there are an even number of entries off the main diagonal assigned 0. Hence the number of entries on the main diagonal which are assigned 0 is also even. This shows that n is even.

On the other hand, for $n = 2k$, by tiling with the following 4×4 matrix

$$T = \begin{pmatrix} 0 & 1 & 0 & 1 \\ 1 & 0 & 1 & 0 \\ 0 & 1 & 1 & 0 \\ 1 & 0 & 0 & 1 \end{pmatrix},$$

we construct a $2n \times 2n$ symmetric matrix

$$\begin{pmatrix} T & T & \cdots & T \\ T & T & \cdots & T \\ \vdots & \vdots & & \vdots \\ T & T & \cdots & T \end{pmatrix}.$$

This gives the required assignment of B.

【Score Situation】 This particular problem saw the following distribution of scores among contestants: 72 contestants scored 7 points, 7 contestants scored 6 points, 24 contestants scored 5 points, 25 contestants scored 4 points, 19 contestants scored 3 points, 13 contestants scored 2 points, 16 contestants scored 1 point, and 92 contestants scored 0 point. The average score of this problem is 3.228, indicating that it was relatively straightforward.

Among the top five teams in the team scores, the scores of this problem are as follows: the Soviet Union team scored 42 points (with a total team score of 217 points), the China team scored 41 points (with a total team score of 201 points), the Romania team scored 33 points (with a total team score of 201 points), the Germany team scored 42 points (with a total team score of 174 points), and the Vietnam team scored 32 points (with a total team score of 166 points).

The gold medal cutoff for this IMO was set at 32 points (with 17 contestants earning gold medals), the silver medal cutoff was 23 points (with 48 contestants earning silver medals), and the bronze medal cutoff was 14 points (with 65 contestants earning bronze medals).

In this IMO, a total of five contestants achieved a perfect score of 42 points.

Problem 1.8 (IMO 30-1, proposed by the Philippines). Prove that the set $\{1, 2, \ldots, 1989\}$ can be expressed as the disjoint union of subsets $A_i (i = 1, 2, \ldots, 117)$ such that:

(i) each A_i contains 17 elements;
(ii) the sum of all elements in each A_i is the same.

Proof. Note that $1989 = 17 \times 117$ and divide the set $\{1, 2, \ldots, 1989\}$ into 17 subsets: $\{1, 2, \ldots, 117\}$, $\{117 + 1, 117 + 2, \ldots, 117 + 117\}$, $\{2 \times 117 + 1, 2 \times 117 + 2, \ldots, 2 \times 117 + 117\}, \ldots, \{16 \times 117 + 1, 16 \times 117 + 2, \ldots, 16 \times 117 + 117\}$.

Starting from the fourth subset, place the elements of the $2k$th ($2 \leq k \leq 8$) subset in descending order into $A_1, A_2, \ldots, A_{117}$; place the elements of the $(2k + 1)$th ($2 \leq k \leq 8$) subset in ascending order into $A_1, A_2, \ldots, A_{117}$.

This arrangement ensures that the sums of the 14 elements already placed in A_i are equal.

Place the elements $1, 2, \ldots, 351$ into $A_1, A_2, \ldots, A_{117}$. Arrange the multiple of 3 in descending order as $351, 348, \ldots, 6, 3$ and divide the others into two sequences: $1, 2, 4, 5, \ldots, 175$ and $176, 178, \ldots, 349, 350$, each sequence containing 117 elements, and then arrange them as follows:

$$
\begin{array}{ccccccc}
351, & 348, & 345, & \ldots, & 6, & 3, \\
1, & 2, & 4, & \ldots, & 173, & 175, \\
176, & 178, & 179, & \ldots, & 349, & 350.
\end{array}
$$

Place the first column into A_1, the second column into $A_2, \ldots$, the 117th column into A_{117}. Therefore, it is evident that the sums of the 17 elements in each A_i are equal.

【Score Situation】 This particular problem saw the following distribution of scores among contestants: 91 contestants scored 7 points, 15 contestants scored 6 points, 8 contestants scored 5 points, 4 contestants scored 4 points, 4 contestants scored 3 points, 29 contestants scored 2 points, 31 contestants scored 1 point, and 109 contestants scored 0 point. The average score of this problem is 3.038, indicating that it was relatively straightforward.

Among the top five teams in the team scores, the scores of this problem are as follows: the China team scored 40 points (with a total team score of 237 points), the Romania team scored 37 points (with a total team score of 223 points), the Soviet Union team scored 35 points (with a total team score of 217 points), the German Democratic Republic team scored 36 points (with a total team score of 216 points), and the United States team scored 35 points (with a total team score of 207 points).

The gold medal cutoff for this IMO was set at 38 points (with 20 contestants earning gold medals), the silver medal cutoff was 30 points (with 55 contestants earning silver medals), and the bronze medal cutoff was 18 points (with 72 contestants earning bronze medals).

In this IMO, a total of 10 contestants achieved a perfect score of 42 points.

Problem 1.9 (IMO 30-3, proposed by Denmark). Let n and k be positive integers and let S be a set of n points in a plane such that:

(i) no three points of S are collinear;

(ii) for any point P of S there are at least k points of S equidistant from P.

Prove that $k < \frac{1}{2} + \sqrt{2n}$.

Proof. By condition (ii), for each point P of S, there are at least C_k^2 unordered pairs (A, B) of points in S such that P is on the perpendicular bisector of segment AB. Summing over P, we have counted at least nC_k^2 such unordered pairs.

On the other hand, by condition (i), any unordered pair can only be counted at most twice; otherwise there are three points of S on its perpendicular bisector. Since the number of unordered pairs of S is C_n^2, we have

$$nC_k^2 \leq 2C_n^2,$$

i.e.,
$$nk(k-1) \leq 2n(n-1), \quad k^2 - k - 2(n-1) \leq 0,$$

$$k \leq \frac{1}{2}(1 + \sqrt{8n-7}) < \frac{1}{2}(1 + \sqrt{8n}) = \frac{1}{2} + \sqrt{2n}.$$

The proof is completed.

【Score Situation】 This particular problem saw the following distribution of scores among contestants: 32 contestants scored 7 points, 3 contestants scored 6 points, 7 contestants scored 5 points, 2 contestants scored 4 points, 13 contestants scored 3 points, 19 contestants scored 2 points, 27 contestants scored 1 point, and 188 contestants scored 0 point. The average score of this problem is 1.337, indicating that it was relatively challenging.

Among the top five teams in the team scores, the scores of this problem are as follows: the China team scored 37 points (with a total team score of 237 points), the Romania team scored 26 points (with a total team score of 223 points), the Soviet Union team scored 26 points (with a total team score of 217 points), the German Democratic Republic team scored 29 points (with a total team score of 216 points), and the United States team scored 14 points (with a total team score of 207 points).

The gold medal cutoff for this IMO was set at 38 points (with 20 contestants earning gold medals), the silver medal cutoff was 30 points (with 55 contestants earning silver medals), and the bronze medal cutoff was 18 points (with 72 contestants earning bronze medals).

In this IMO, a total of 10 contestants achieved a perfect score of 42 points.

Problem 1.10 (IMO 30-6, proposed by Poland). A permutation $(x_1, x_2, \ldots, x_{2n})$ of the set $\{1, 2, \ldots, 2n\}$, where n is a positive integer, is said to have property P if $|x_i - x_{i+1}| = n$ for at least one i in $\{1, 2, \ldots, 2n-1\}$. Show that, for each n, there are more permutations with property P than without.

Proof 1. Let $N_k(1 \leq k \leq n)$ be the set of permutations $(x_1, x_2, \ldots, x_{2n})$ such that k and $k + n$ are adjacent. By the Inclusion-Exclusion Principle, the number of permutations with property P, denoted by m, satisfies

$$m \geq \sum_{k=1}^{n} |N_k| - \sum_{1 \leq k < l \leq n} \left| N_k \bigcap N_l \right|.$$

Since $|N_k| = 2 \times (2n - 1)!$, we have $|N_k \bigcap N_l| = 2^2 \times (2n - 2)!$ (We may group up k and $k + n$, treat them as a single number, and group up l and $l + n$ as well. Now there are $(2n - 2)!$ permutations for $2n - 2$ numbers. By swapping k and $k+n$, or l and $l+n$, each permutation yields 2^2 permutations in $N_k \bigcap N_l$), and thus

$$m \geq 2n(2n - 1)! - 2^2 \times (2n - 2)! C_n^2$$
$$= (2n)! - 2n(n - 1)(2n - 2)!$$
$$= 2n \cdot (2n - 2)! \cdot n$$
$$> \frac{1}{2} \cdot (2n)!.$$

Proof 2. We call i and $i + n$ friends for $1 \leq i \leq n$. The property P is the same as saying at least one pair of friends are neighbors in the permutation.

For any permutation $(x_1, x_2, \ldots, x_{2n})$ without property P, there exists a unique $3 \leq j \leq 2n$ such that $|x_1 - x_j| = n$. We map this permutation to the permutation $(x_2, \ldots, x_{j-1}, x_1, x_j, x_{j+1}, \ldots, x_{2n})$, which moves x_1 immediately prior to x_j. This is a permutation with property P, and there exists exactly one pair of friends which are neighbors, not on the first two terms. This correspondence is clearly injective. However, there exists a permutation with property P whose first two terms are friends. So, there are more permutations with property P than without.

【Score Situation】This particular problem saw the following distribution of scores among contestants: 69 contestants scored 7 points, 11 contestants scored 6 points, 15 contestants scored 5 points, 11 contestants scored 4 points, 6 contestants scored 3 points, 13 contestants scored 2 points, 23 contestants scored 1 point, and 143 contestants scored 0 point. The average score of this problem is 2.526, indicating that it had a certain level of difficulty.

Among the top five teams in the team scores, the scores of this problem are as follows: the China team scored 35 points (with a total team score of 237 points), the Romania team scored 35 points (with a total team score of 223 points), the Soviet Union team scored 33 points (with a total team score of 217 points), the German Democratic Republic team scored 33 points (with a total team score of 216 points), and the United States team scored 38 points (with a total team score of 207 points).

The gold medal cutoff for this IMO was set at 38 points (with 20 contestants earning gold medals), the silver medal cutoff was 30 points (with 55 contestants earning silver medals), and the bronze medal cutoff was 18 points (with 72 contestants earning bronze medals).

In this IMO, a total of 10 contestants achieved a perfect score of 42 points.

Problem 1.11 (IMO 33-5, proposed by Italy). Let S be a finite set of points in a three-dimensional space. Let S_x, S_y, and S_z be the sets consisting of the orthogonal projections of the points of S onto the yz-plane, zx-plane, and xy-plane, respectively. Prove that

$$|S|^2 \leq |S_x| \cdot |S_y| \cdot |S_z|,$$

where $|A|$ denotes the number of elements in a finite set A. (*Note:* The orthogonal projection of a point onto a plane is the foot of the perpendicular from that point to the plane.)

Proof. The points of S are lying on $|S_z|$ lines parallel to z-axis. Denote the number of points on each of these lines by $t_{ij}((i,j,0) \in S_z)$. By the Cauchy–Schwarz inequality,

$$|S|^2 = \left(\sum_{(i,j,0) \in S_z} t_{ij} \right)^2 \leq \sum_{(i,j,0) \in S_z} 1^2 \cdot \sum_{(i,j,0) \in S_z} t_{ij}^2$$

$$= |S_z| \sum_{(i,j,0) \in S_z} t_{ij}^2.$$

Let $u_i = |\{(i,0,z) \in S_y\}|$ and $v_j = |\{(0,j,k) \in S_x\}|$, then $u_i \geq t_{ij}$, $v_j \geq t_{ij}$, $|S_x| = \sum u_i$, and $|S_y| = \sum v_j$. Therefore,

$$|S_x| \cdot |S_y| = \sum_{(i,j,0) \in S_z} u_i v_j \geq \sum_{(i,j,0) \in S_z} t_{ij}^2.$$

Hence $|S|^2 \leq |S_x| \cdot |S_y| \cdot |S_z|$.

【Score Situation】This particular problem saw the following distribution of scores among contestants: 22 contestants scored 7 points, 2 contestants scored 6 points, 2 contestants scored 5 points, no contestant scored 4 points, 7 contestants scored 3 points, 3 contestants scored 2 points, 62 contestants scored 1 point, and 252 contestants scored 0 point. The average score of this problem is 0.757, indicating that it was extremely difficult.

Among the top five teams in the team scores, the scores of this problem are as follows: the China team scored 35 points (with a total team score of 240 points), the United States team scored 11 points (with a total team score of 181 points), the Romania team scored 9 points (with a total team score of 177 points), the Commonwealth of Independent States team scored 15 points (with a total team score of 176 points), and the United Kingdom team scored 21 points (with a total team score of 168 points).

The gold medal cutoff for this IMO was set at 32 points (with 26 contestants earning gold medals), the silver medal cutoff was 24 points (with 55 contestants earning silver medals), and the bronze medal cutoff was 14 points (with 74 contestants earning bronze medals).

In this IMO, a total of four contestants achieved a perfect score of 42 points.

Problem 1.12 (IMO 39-2, proposed by India). In a competition, there are a contestants and b judges, where $b \geq 3$ is an odd integer. Each judge rates each contestant as either "pass" or "fail." Suppose k is a number such that for any two judges, their ratings coincide for at most k contestants. Prove that

$$\frac{k}{a} \geq \frac{b-1}{2b}.$$

Proof. If two judges have the same ratings on a contestant, we say that a coincidence happens. By assumption, any two judges may cause at most k coincidences. Thus, the total number of coincidences S satisfies

$$S \leq k\mathrm{C}_b^2. \tag{1}$$

On the other hand, fix any contestant. Suppose he receives x "passes" and y "fails." Then $x + y = b$. The number of coincidences happened to this contestant is

$$\begin{aligned}
\mathrm{C}_x^2 + \mathrm{C}_y^2 &= \frac{x^2 + y^2 - (x+y)}{2} \\
&= \frac{(x+y)^2 + (x-y)^2 - 2(x+y)}{4} \\
&= \frac{b^2 - 2b + (x-y)^2}{4}.
\end{aligned}$$

Since b is odd, $x - y$ is odd as well. Hence $(x - y)^2 \geq 1$, so

$$C_x^2 + C_y^2 \geq \frac{b^2 - 2b + 1}{4} = \left(\frac{b-1}{2}\right)^2.$$

It follows that the total number of coincidences satisfies

$$S \geq a \cdot \left(\frac{b-1}{2}\right)^2. \tag{2}$$

Combining (1) and (2), we obtain

$$a \cdot \left(\frac{b-1}{2}\right)^2 \leq k \cdot C_b^2,$$

i.e.,

$$a \cdot \left(\frac{b-1}{2}\right)^2 \leq k \cdot \frac{b(b-1)}{2}, \quad \text{and hence} \quad \frac{k}{a} \geq \frac{b-1}{2b}.$$

【Score Situation】This particular problem saw the following distribution of scores among contestants: 141 contestants scored 7 points, 5 contestants scored 6 points, 4 contestants scored 5 points, 5 contestants scored 4 points, 6 contestants scored 3 points, 26 contestants scored 2 points, 19 contestants scored 1 point, and 213 contestants scored 0 point. The average score of this problem is 2.735, indicating that it had a certain level of difficulty.

Among the top five teams in the team scores, the scores of this problem are as follows: the Iran team scored 42 points (with a total team score of 211 points), the Bulgaria team scored 36 points (with a total team score of 195 points), the Hungary team scored 42 points (with a total team score of 186 points), the United States team scored 35 points (with a total team score of 186 points), and the Chinese Taipei team scored 27 points (with a total team score of 184 points).

The gold medal cutoff for this IMO was set at 31 points (with 37 contestants earning gold medals), the silver medal cutoff was 24 points (with 66 contestants earning silver medals), and the bronze medal cutoff was 14 points (with 102 contestants earning bronze medals).

In this IMO, only one contestant achieved a perfect score of 42 points, namely Omid Amini from Iran.

Problem 1.13 (IMO 43-1, proposed by Colombia).

Suppose T is the set of all (x, y) with x and y non-negative integers such that $x + y < n$. Each element of T is colored red or blue, so that if (x, y) is red and $x' \leq x$, $y' \leq y$, then (x', y') is also red. A type 1 subset of T has n blue elements with different first members and a type 2 subset of T has n blue elements with different second members. Show that there are the same number of type 1 and type 2 subsets.

Proof 1. Let m denote the number of red points of T, t denote the number of type 1 subsets, and s denote the number of type 2 subsets. We induct on m for $0 \le m \le \frac{n(n+1)}{2}$.

For $m = 0$, we have $s = t = n!$ by the principle of multiplication. The statement is true.

Assume that the statement is true for $m = k$, where $0 \le k < \frac{n(n+1)}{2}$.

For $m = k + 1$, let $P(x_0, y_0)$ be a red point for which $x_0 + y_0$ attains a maximum. Change the color of P to blue. Since there is no other red point (x, y) with $x \ge x_0$ and $y \ge y_0$, the colors of points satisfy the assumption. Denote this colored set by T' and the corresponding new values of t, s by t', s'.

Assume that there are a_i blue points on the line $x = i$ in T' and b_j blue points on the line $y = j$ with $0 \le i, j \le n - 1$. Clearly $t' = \prod_{i=0}^{n-1} a_i$ and $s' = \prod_{j=0}^{n-1} b_j$. We have $t' = s'$ by the inductive hypothesis. Hence $\prod_{i=0}^{n-1} a_i = \prod_{j=0}^{n-1} b_j$.

For T, we have

$$t = \left(\prod_{\substack{i=0 \\ i \ne x_0}}^{n-1} a_i \right)(a_{x_0} - 1) \quad \text{and} \quad s = \left(\prod_{\substack{j=0 \\ j \ne y_0}}^{n-1} b_j \right)(b_{y_0} - 1).$$

On the line $x = x_0$, there are $n - x_0 - y_0$ points in T with the second coordinate greater than or equal to y_0, and hence $a_{x_0} = n - x_0 - y_0$. Similarly, $b_{y_0} = n - x_0 - y_0$. Thus $a_{x_0} = b_{y_0}$. Consequently,

$$\left(\prod_{\substack{i=0 \\ i \ne x_0}}^{n-1} a_i \right)(a_{x_0} - 1) = \left(\prod_{\substack{j=0 \\ j \ne y_0}}^{n-1} b_j \right)(b_{y_0} - 1),$$

i.e., $t = s$. The statement also holds for $m = k + 1$.

Hence, the statement holds for every $0 \le m \le \frac{n(n+1)}{2}$ by mathematical induction.

Proof 2. For $0 \leq i, j \leq n - 1$, let a_i be the number of blue points with first coordinate i and b_j the number of blue points with second coordinate j. We need to prove that

$$a_0 a_1 \ldots a_{n-1} = b_0 b_1 \ldots b_{n-1}.$$

It suffices to show that $(a_0, a_1, \ldots, a_{n-1})$ is a permutation of $(b_0, b_1, \ldots, b_{n-1})$.

We induct on n to prove the above statement. It is trivial for $n = 1$. Assume that the statement holds for $n \leq k$, where k is a positive integer. For $n = k + 1$, we distinguish two cases.

Case 1: The points (x, y) with $x + y = n - 1 = k$ are all blue, as shown in the left figure of Figure 1.2. By deleting these blue points, we reduce the configuration to the case of $n = k$, where the columns contain $a_0 - 1, a_1 - 1, \ldots, a_{n-2} - 1$ blue points and the rows contain $b_0 - 1, b_1 - 1, \ldots, b_{n-2} - 1$ blue points, respectively. It follows from the inductive hypothesis that $(a_0 - 1, a_1 - 1, \ldots, a_{n-2} - 1)$ is a permutation of $(b_0 - 1, b_1 - 1, \ldots, b_{n-2} - 1)$. Note that $a_{n-1} = b_{n-1} = 1$, and so the statement holds for $n = k + 1$ in this case.

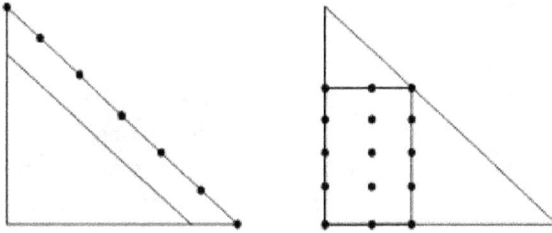

Figure 1.2 Two situations for $n = k + 1$

Case 2: Some point $(m, n - m - 1) = (m, k - m)$ is red, as shown in the right figure of Figure 1.2. By assumption, the points in the rectangle defined by $0 \leq x \leq m$ and $0 \leq y \leq k - m$ are all red. The blue points are separated in two smaller isosceles right triangles. By the inductive hypothesis, $(a_0, a_1, \ldots, a_{m-1})$ is a permutation of $(b_{n-m}, b_{n-m+1}, \ldots, b_{n-1})$, and similarly $(a_{m+1}, a_{m+2}, \ldots, a_{n-1})$ is a permutation of $(b_0, b_1, \ldots, b_{n-m-2})$. Since $a_m = b_{n-m-1} = 0$, the statement also holds for $n = k + 1$ in this case.

The proof is completed by mathematical induction.

【Score Situation】 This particular problem saw the following distribution of scores among contestants: 155 contestants scored 7 points, 61 contestants scored 6 points, 15 contestants scored 5 points, 8 contestants scored 4 points, 11 contestants scored 3 points, 11 contestants scored 2 points, 39 contestants scored 1 point, and 179 contestants scored 0 point. The average score of this problem is 3.449, indicating that it was relatively straightforward.

Among the top five teams in the team scores, the scores of this problem are as follows: the China team scored 41 points (with a total team score of 212 points), the Russia team scored 42 points (with a total team score of 204 points), the United States team scored 38 points (with a total team score of 171 points), the Bulgaria team scored 32 points (with a total team score of 167 points), and the Vietnam team scored 32 points (with a total team score of 166 points).

The gold medal cutoff for this IMO was set at 29 points (with 39 contestants earning gold medals), the silver medal cutoff was 23 points (with 73 contestants earning silver medals), and the bronze medal cutoff was 14 points (with 120 contestants earning bronze medals).

In this IMO, only three contestants achieved a perfect score of 42 points, namely Yunhao Fu and Botong Wang from China, and Andrei Khaliavine from Russia.

Problem 1.14 (IMO 44-1, proposed by Brazil). Let S be the set $\{1, 2, \ldots, 1000000\}$. Show that for any subset A of S with 101 elements we can find 100 distinct elements x_i of S such that the sets $A_i = \{a + x_i \mid a \in A\}$ are all pairwise disjoint.

Proof. Consider the set $D = \{x - y \mid x, y \in A\}$. There are at most $101 \times 100 + 1 = 10101$ elements in D. Two sets A_i and A_j have a nonempty intersection if and only if $x_i - x_j$ is in D. So, we need to choose the 100 elements in such a way that the difference of any two elements is not in D.

Now select these elements by induction. Assume that k elements, $k \le 99$, are already chosen. We need to select an $x \in S$ such that k chosen elements are not from the set $x + D$, where $x + D = \{x + y \mid y \in D\}$. Thus, after k elements are chosen, at most $10101k \le 999999$ elements in S are forbidden. Hence, we can select one more element.

Remark. The size $|S| = 10^6$ is unnecessarily large. The following statement is true:

If A is a k-element subset of $S = \{1, 2, \ldots, n\}$ and m is a positive integer such that

$$n > (m - 1)\left(\binom{k}{2} + 1\right),$$

then there exist $t_1, t_2, \ldots, t_m \in S$ such that the sets $A_i = \{x + t_j | x \in A\}$ for $j = 1, 2, \ldots, m$ are pairwise disjoint.

【Score Situation】 This particular problem saw the following distribution of scores among contestants: 180 contestants scored 7 points, 1 contestant scored 6 points, 24 contestants scored 5 points, no contestant scored 4 points, no contestant scored 3 points, 111 contestants scored 2 points, 18 contestants scored 1 point, and 123 contestants scored 0 point. The average score of this problem is 3.558, indicating that it was relatively straightforward.

Among the top five teams in the team scores, the scores of this problem are as follows: the Bulgaria team scored 37 points (with a total team score of 227 points), the China team scored 42 points (with a total team score of 211 points), the United States team scored 42 points (with a total team score of 188 points), the Vietnam team scored 26 points (with a total team score of 172 points), and the Russia team scored 37 points (with a total team score of 167 points).

The gold medal cutoff for this IMO was set at 29 points (with 37 contestants earning gold medals), the silver medal cutoff was 19 points (with 69 contestants earning silver medals), and the bronze medal cutoff was 13 points (with 104 contestants earning bronze medals).

In this IMO, only three contestants achieved a perfect score of 42 points, namely Bảo Lê Hùng Việt and Trọng Cảnh Nguyễn from Vietnam, and Yunhao Fu from China.

Problem 1.15 (IMO 49-5, proposed by France). Let n and k be positive integers with $k \geq n$ and $k - n$ an even number. Let $2n$ lamps labeled $1, 2, \ldots, 2n$ be given, each of which can be either *on* or *off*. Initially all the lamps are *off*. We consider sequences of steps: at each step one of the lamps is switched (from *on* to *off* or from *off* to *on*).

Let N be the number of such sequences consisting of k steps and resulting in the state where lamps 1 through n are all *on*, and lamps $n+1$ through $2n$ are all *off*.

Let M be the number of such sequences consisting of k steps, resulting in the state where lamps 1 through n are all *on*, and lamps $n+1$ through $2n$ are all *off*, but where none of the lamps $n + 1$ through $2n$ is ever switched *on*. Determine the ratio $\frac{N}{M}$.

Solution (By Ruixiang Zhang). The ratio is 2^{k-n}.

Lemma. For any positive integer t, call a t-element array $(a_1, a_2, \ldots, a_t)$ which is made of 0, 1 ($a_1, a_2, \ldots, a_t \in \{0, 1\}$) "good" if there are odd 0s in it. Prove that there are 2^{t-1} "good" arrays.

Proof of the Lemma. In fact, for same $a_1, a_2, \ldots, a_{t-1}$, when a_t is 0 or 1, the parity of 0s in these two arrays is different, so only one of the array is "good." There are 2^t arrays in all, we can correspond one array to the other, with their difference a_t. Only one of these two arrays is "good." So of all the possible arrays, only half of them are "good." The lemma is proved.

Let A be the set of such sequences consisting of k steps and resulting in the state where lamps 1 through n are all on, and lamps $n + 1$ through $2n$ are all off.

Let B the set of such sequences consisting of k steps, resulting in the state where lamps 1 through n are all on, and lamps $n + 1$ through $2n$ are all off, but where none of the lamps $n + 1$ through $2n$ is ever switched on.

For any b in B, make those a in A correspond to b if "a's elements are the same as b's in $(\bmod\, n)$" (For example, take $n = 2$ and $k = 4$, if $b = (2, 2, 2, 1)$, and then it could correspond to $a = (4, 4, 2, 1)$, $a = (2, 2, 2, 1)$, $a = (2, 4, 4, 1)$ etc.).

Since b is in B, the number of $1, 2, \ldots, n$ in it must be odd; a is in A, the number of $1, 2, \ldots, n$ in it must be odd, and the number of $n+1, n+2, \ldots, 2n$ must be even.

For any $i \in \{1, 2, \ldots, n\}$, if the number of i s in b is b_i, then a only needs to satisfy: for the positions taken by i in b the corresponding positions taken by i or $n + i$ in a, and the number of i s is odd (thus the number of $n + i$ is even). By the lemma and the Principle of Multiplication, there are $\prod_{i=1}^{n} 2^{b_i - 1} = 2^{k-n}$ a s that correspond to b, but only one of b (which lets every position of a be the remainder of n) in B corresponds to each a in A.

Therefore $|A| = 2^{k-n}|B|$, i.e., $N = 2^{k-n}M$.

Obviously $M \neq 0$ (because the sequence $(1, 2, \ldots, n - 1, n, n, \ldots, n) \in B$), so $\frac{N}{M} = 2^{k-n}$.

【Score Situation】 This particular problem saw the following distribution of scores among contestants: 132 contestants scored 7 points, 1 contestant scored 6 points, 4 contestants scored 5 points, 1 contestant scored 4 points, 11 contestants scored 3 points, 33 contestants scored 2 points, 58 contestants scored 1 point, and 295 contestants scored 0 point. The average score of this problem is 2.077, indicating that it had a certain level of difficulty.

Among the top five teams in the team scores, the scores of this problem are as follows: the China team scored 35 points (with a total team score of 217 points), the Russia team scored 42 points (with a total team score of 199 points), the United States team scored 42 points (with a total team score of 190 points), the South Korea team scored 35 points (with a total team score of 188 points), and the Iran team scored 37 points (with a total team score of 181 points).

The gold medal cutoff for this IMO was set at 31 points (with 47 contestants earning gold medals), the silver medal cutoff was 22 points (with 100 contestants earning silver medals), and the bronze medal cutoff was 15 points (with 120 contestants earning bronze medals).

In this IMO, only three contestants achieved a perfect score of 42 points, namely Xiaosheng Mu and Dongyi Wei from China, and Alex Zhai from the United States.

Problem 1.16 (IMO 50-6, proposed by Russia). Let $a_1, a_2, \ldots, a_n$ be distinct positive integers and let M be a set of $n - 1$ positive integers not containing $s = a_1 + a_2 + \cdots + a_n$. A grasshopper is to jump along the real axis, starting at the point 0 and making n jumps to the right with lengths $a_1, a_2, \ldots, a_n$ in some order. Prove that the order can be chosen in such a way that the grasshopper never lands on any point in M.

Proof. We proceed by induction on n. When $n = 1$, clearly M is empty, and the result is obvious.

When $n = 2$, we see that M does not contain at least one of a_1 and a_2, and the grasshopper can first jump the number that is not in M, then the other number.

Let $m \geq 3$ and assume that the result is true for any $n < m$ with $n \in \mathbf{N}_+$. We shall prove that the result also holds for $n = m$. Suppose on the contrary that there exist m distinct positive integers $a_1, a_2, \ldots, a_m$ and a set M of $m - 1$ numbers such that the grasshopper cannot finish the jumping as required in the problem.

Suppose that the grasshopper can jump at most k steps to the right with lengths of distinct numbers in $a_1, a_2, \ldots, a_m$ such that the landing points are never in M. Since M contains only $m - 1$ numbers, the grasshopper can always take its first step. If the grasshopper can jump $m - 1$ steps, it can also jump the last step since M does not contain $a_1 + a_2 + \cdots + a_n$, and therefore $1 \leq k \leq m - 2$.

We choose such k steps with a minimum total length (if there is a tie, just choose any one that attains the minimum), denote the lengths of these k steps by $b_1, b_2, \ldots, b_k$, and let $b_{k+1}, b_{k+2}, \ldots, b_m$ be the remaining numbers in $a_1, a_2, \ldots, a_m$ in increasing order. Clearly,

$$\{a_1, a_2, \ldots, a_m\} = \{b_1, b_2, \ldots, b_m\}.$$

By assumption, $b_1 + b_2 + \cdots + b_k + b_j$ belongs to M for any $k+1 \leq j \leq m$. Thus, there are at least $m - k$ elements of M that are greater than or equal to $b_1 + b_2 + \cdots + b_{k+1}$, and hence there are at most $k - 1$ elements of M that are less than $b_1 + b_2 + \cdots + b_{k+1}$.

Let $A = \{b_i | 1 \le i \le k + 1, i \in \mathbf{Z}, b_1 + b_2 + \cdots + b_{k+1} - b_i \notin M\}$.

Then $b_{k+1} \in A$. For any $b_i \in A$, if we remove b_i from $b_1, b_2, \ldots, b_{k+1}$, then the sum of the remaining k numbers is not in M, and since

$$\left| M \bigcap \{1, 2, \ldots, b_1 + b_2 + \cdots + b_{k+1} - b_i\} \right| \le k - 1,$$

and $k < m$, by the inductive hypothesis there exists an ordering $c_1, c_2, \ldots, c_k$ of these remaining numbers such that

$$c_1, c_1 + c_2, \ldots, c_1 + c_2 + \cdots + c_k$$

are not in M. Thus, $c_1, c_2, \ldots, c_k$ is also a possible length of k jumps, and by the choice of $b_1, b_2, \ldots, b_k$ we have

$$b_1 + b_2 + \cdots + b_k \le c_1 + c_2 + \cdots + c_k,$$

i.e., $b_i \le b_{k+1}$ for any $b_i \in A$.

Let $A = \{x_1, x_2, \ldots, x_t\}$, where $x_1 < x_2 < \cdots < x_t = b_{k+1}$. By the definition of A, there are at least $k + 1 - t$ numbers of M that are less than $b_1 + b_2 + \cdots + b_{k+1}$. Since $b_1 + b_2 + \cdots + b_k + b_j$ belongs to M for any $k + 1 \le j \le m$, there are at least $m - k$ numbers of M in the interval

$$[b_1 + b_2 + \cdots + b_{k+1}, b_1 + b_2 + \cdots + b_k + b_m].$$

On the other hand, for any $x_i \in A$, the number

$$b_1 + b_2 + \cdots + b_{k+1} - x_i + b_m$$

belongs to M (otherwise, b_m is different from $b_1, b_2, \ldots, b_{k+1}$ since $k \le m - 2$, and the grasshopper can take $k + 1$ jumps).

For $1 \le i \le t - 1$, the numbers

$$b_1 + b_2 + \cdots + b_{k+1} - x_i + b_m$$

are pairwise distinct and greater than $b_1 + b_2 + \cdots + b_k + b_m$, and hence there are at least $t - 1$ numbers of M greater than $b_1 + b_2 + \cdots + b_k + b_m$. Summing up, M contains at least

$$(k + 1 - t) + (m - k) + (t - 1) = m$$

numbers, which contradicts the fact that M contains only $m - 1$ numbers.

Thus, our assumption is false, and the result also holds for $n = m$. By the method of mathematical induction, the result holds for every positive integer n.

【Score Situation】 This particular problem saw the following distribution of scores among contestants: 3 contestants scored 7 points, 1 contestant scored 6 points, 2 contestants scored 5 points, 6 contestants scored 4 points, 10 contestants scored 3 points, 1 contestant scored 2 points, 2 contestants scored 1 point, and 540 contestants scored 0 point. The average score of this problem is 0.168, indicating that it was extremely difficult.

Among the top five teams in the team scores, the scores of this problem are as follows: the China team scored 11 points (with a total team score of 221 points), the Japan team scored 19 points (with a total team score of 212 points), the Russia team scored 7 points (with a total team score of 203 points), the South Korea team scored 0 point (with a total team score of 188 points), and the North Korea team scored 7 points (with a total team score of 183 points).

The gold medal cutoff for this IMO was set at 32 points (with 49 contestants earning gold medals), the silver medal cutoff was 24 points (with 98 contestants earning silver medals), and the bronze medal cutoff was 14 points (with 135 contestants earning bronze medals).

In this IMO, only two contestants achieved a perfect score of 42 points, namely Dongyi Wei from China and Makoto Soejima from Japan.

Problem 1.17 (IMO 52-1, proposed by Mexico). Given any set $A = \{a_1, a_2, a_3, a_4\}$ of four distinct positive integers, we denote the sum $a_1 + a_2 + a_3 + a_4$ by s_A. Let n_A denote the number of pairs (i, j) with $1 \leq i < j \leq 4$ for which $a_i + a_j$ divides s_A. Find all sets A of four distinct positive integers which achieve the largest possible value of n_A.

Solution. Let $A = \{a_1, a_2, a_3, a_4\}$ be the set of four positive integers with $a_1 < a_2 < a_3 < a_4$. Since

$$\frac{1}{2}s_A = \frac{1}{2}(a_1 + a_2 + a_3 + a_4) < a_2 + a_4 < a_3 + a_4 < s_A,$$

$a_2 + a_4$ and $a_3 + a_4$ do not divide s_A. Therefore,

$$n_A \leq C_4^2 - 2 = 4.$$

On the other hand, if $A = \{1, 5, 7, 11\}$, then $n_A = 4$, so the largest possible value of n_A is 4.

Next, we will find all sets A of four positive integers with $n_A = 4$.
First, $a_2 + a_4$ and $a_3 + a_4$ do not divide s_A, and

$$\frac{1}{2}s_A \leq \max\{a_1 + a_4, a_2 + a_3\} < s_A.$$

Thus, $\frac{1}{2}s_A = \max\{a_1 + a_4, a_2 + a_3\}$, and then $a_1 + a_4 = a_2 + a_3$.
By $(a_1 + a_3)|s_A$, let $s_A = k(a_1 + a_3)$, where k is a positive integer. And by $a_1 + a_3 < a_2 + a_3$ we know that $k > 2$.

As $2(a_2 + a_3) = s_A = k(a_1 + a_3)$, we have $a_2 = \frac{1}{2}(ka_1 + (k-2)a_3)$, and from

$$a_2 = \frac{1}{2}(ka_1 + (k-2)a_3) < a_3,$$

we obtain $k < 4$. Thus, $k = 3$. Consequently,

$$2(a_2 + a_3) = 2(a_1 + a_4) = 3(a_1 + a_3) = s_A,$$

from which $a_2 = \frac{1}{2}(3a_1 + a_3)$ and $a_4 = \frac{1}{2}(a_1 + 3a_3)$.

From $(a_1 + a_2) | s_A$, by letting $s_A = l(a_1 + a_2)$, we have

$$3(a_1 + a_3) = l\left(a_1 + \frac{1}{2}(3a_1 + a_3)\right),$$

that is, $(6 - l)a_3 = (5l - 6)a_1.$

Since a_1 and a_3 are positive integers and $a_1 < a_3$, we see that $l = 3, 4$, or 5.

If $l = 3$, then $a_2 = a_3$, which is a contradiction.

If $l = 4$, then $a_3 = 7a_1$, and it follows that $a_2 = 5a_1$ and $a_4 = 11a_1$.

If $l = 5$, then $a_3 = 19a_1$, and consequently, $a_2 = 11a_1$ and $a_4 = 29a_1$.

It is easy to verify that, when $l = 4$ and 5, each of $a_1 + a_2$, $a_1 + a_3$, $a_1 + a_4$, and $a_2 + a_3$ divides s_A.

To sum up, all sets A of four distinct positive integers which achieve the largest possible value of $n_A = 4$ are $A = \{a, 5a, 7a, 11a\}$ and $A = \{a, 11a, 19a, 29a\}$, where a is any positive integer.

【Score Situation】 This particular problem saw the following distribution of scores among contestants: 353 contestants scored 7 points, 14 contestants scored 6 points, 17 contestants scored 5 points, 18 contestants scored 4 points, 52 contestants scored 3 points, 63 contestants scored 2 points, 17 contestants scored 1 point, and 29 contestants scored 0 point. The average score of this problem is 5.348, indicating that it was simple.

Among the top five teams in the team scores, the scores of this problem are as follows: the China team scored 42 points (with a total team score of 189 points), the United States team scored 42 points (with a total team score of 184 points), the Singapore team scored 42 points (with a total team score of 179 points), the Russia team scored 42 points (with a total team score of 161 points), and the Thailand team scored 39 points (with a total team score of 160 points).

The gold medal cutoff for this IMO was set at 28 points (with 54 contestants earning gold medals), the silver medal cutoff was 22 points (with 90 contestants earning silver medals), and the bronze medal cutoff was 16 points (with 137 contestants earning bronze medals).

In this IMO, only one contestant achieved a perfect score of 42 points, namely Lisa Sauermann from Germany.

Problem 1.18 (IMO 52-4, proposed by Iran). Let $n > 0$ be an integer. We are given a balance and n weights of $2^0, 2^1, \ldots, 2^{n-1}$. We are to place each of the n weights on the balance, one after another, in such a way that the right pan is never heavier than the left pan. At each step we choose one of the weights that has not yet been placed on the balance and place it on either the left pan or the right pan, until all the weights have been placed. Determine the number of ways in which this can be done.

Solution. The number of ways is $(2n-1)!! = 1 \times 3 \times 5 \times \cdots \times (2n-1)$.

We prove the conclusion by induction on n. The case $n = 1$ is clear: put one weight on the left pan. Hence, there is only one way.

Suppose that in the case $n = k$, the number of ways is $(2k-1)!!$.

Then for the case $n = k+1$, without affecting the result, multiply both weights by $\frac{1}{2}$, and then $k+1$ weights are $\frac{1}{2}, 1, 2, \ldots, 2^{k-1}$. Since for any positive integer r,

$$2^r > 2^{r-1} + 2^{r-2} + \cdots + 1 + \frac{1}{2} \geq \sum_{i=-1}^{r-1} \pm 2^i,$$

the heavier pan is where the heaviest weight is. Therefore, the heaviest weight on balance must be on the left pan. In the following, consider the position of the weight $\frac{1}{2}$ in the process.

(1) If we take weight $\frac{1}{2}$ first, then it must be put on the left pan. Then the remaining k weights have $(2k-1)!!$ ways to do.

(2) If we take weight $\frac{1}{2}$ at step $t = 2, 3, \ldots, k+1$, then we can put it either on the left pan or on the right pan because weight $\frac{1}{2}$ is the lightest one, so the number of ways is $2k \times (2k-1)!!$.

To sum up, when $n = k+1$, there are totally

$$(2k-1)!! + 2k \times (2k-1)!! = (1+2k)(2k-1)!! = (2k+1)!!$$

ways, which completes the induction proof.

【Score Situation】 This particular problem saw the following distribution of scores among contestants: 267 contestants scored 7 points, 20 contestants scored 6 points, 8 contestants scored 5 points, 8 contestants scored 4 points, 16 contestants scored 3 points, 31 contestants scored 2 points, 120 contestants scored 1 point, and 93 contestants scored 0 point. The average score of this problem is 4.069, indicating that it was simple.

Among the top five teams in the team scores, the scores of this problem are as follows: the China team scored 42 points (with a total team score of 189 points), the United States team scored 42 points (with a total team score of 184 points), the Singapore team scored 42 points (with a total team score of 179 points), the Russia team scored 40 points (with a total team score of 161 points), and the Thailand team scored 42 points (with a total team score of 160 points).

The gold medal cutoff for this IMO was set at 28 points (with 54 contestants earning gold medals), the silver medal cutoff was 22 points (with 90 contestants earning silver medals), and the bronze medal cutoff was 16 points (with 137 contestants earning bronze medals).

In this IMO, only one contestant achieved a perfect score of 42 points, namely Lisa Sauermann from Germany.

Problem 1.19 (IMO 54-6, proposed by Russia). Let $n \geq 3$ be an integer and consider a circle with $n+1$ equally spaced points marked on it. Consider all labelings of these points with the numbers $0, 1, \ldots, n$ such that each label is used exactly once; two such labelings are the same if one can be obtained from the other by a rotation of the circle. A labeling is called *beautiful* if, for any four labels $a < b < c < d$ with $a + d = b + c$, the chord joining the points labeled a and d does not intersect the chord joining the points labeled b and c.

Let M be the number of beautiful labelings, and let N be the number of ordered pairs (x, y) of positive integers such that $x + y \leq n$ and $\gcd(x, y) = 1$. Prove that

$$M = N + 1.$$

Proof. Note that the distance between marked points does not matter. The intersection of the chords only depends on the order of the points. For a circulation of permutation of $[0, n] = \{0, 1, \ldots, n\}$, we call a chord (x, y) a k-chord if $x + y = k$. If $x = y$, then the chord degenerates. We call three disjoint chords in order if a chord parts other two chords, e.g., in Figure 1.3, chords A, B, and C are in order but chords B, C, and D are not. We call $m \geq 3$ disjoint chords in order if any three chords are in order.

Lemma. In a beautiful labeling, all k-chords are in order for any integer k.

Proof of the Lemma. We prove it by induction on n. The lemma is trivial for $n \leq 3$. For $n \geq 4$, suppose that there were a beautiful labeling S such that three k-chords A, B, and C were not in order. If n is not the end point of chords A, B, and C, then we can delete the point n and obtain a beautiful labeling $S \backslash \{n\}$ of $[0, n-1]$.

By the hypothesis of induction, chords A, B, and C are in order. Similarly, if 0 is not the end point of chords A, B, and C, then we can delete

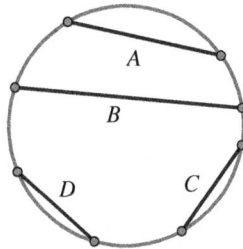

Figure 1.3 An Example for Chords in Oder

the point 0 and minus each labeling by 1, so we obtain a beautiful labeling $S\backslash\{0\}$. By the hypothesis of induction, chords A, B, and C are in order, which is a contradiction.

Thus, 0 and n must appear among the end points of n-chords of A, B, and C. Suppose that chords $(0, x)$ and (y, n) are among A, B, and C. Then $n \geq 0 + x = k = n + y$. Thus $x = n$ and $y = 0$. That is, $(0, n)$ is one of n-chords of A, B, and C. Without loss of generality, suppose that $C = (0, n)$.

Let chord $D = (u, v)$ be adjacent and parallel to chord C, see Figure 1.4, and denote $t = u + v$. If $t = n$, then n-chords A, B, and D are not in order in the beautiful labeling $S\backslash\{0, n\}$, which contradicts the hypothesis of induction. If $t < n$, then t-chord $(0, t)$ does not intersect D. Hence chord C is apart point t and chord D. And n-chord $E = (t, n-t)$ does not intersect chord C. So, points t and $n - t$ are on the same side of C. But chords A, B, and E are not in order, which is a contradiction.

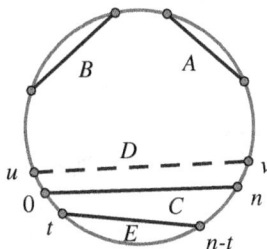

Figure 1.4 An Example for Chord $D = (u, v)$

Lastly, since the mapping of x to $n - x$ conserves the beautifulness of a circular permutation, a t-chord is mapped to $(2n - t)$-chord, that is, $t > n$ is equivalent to $t < n$. Thus, we have proved the auxiliary lemma.

In the following, we prove the original proposition by induction on n. The case of $n = 2$ is trivial. Let $n \geq 3$ and S be a beautiful permutation of $[0, n]$. We obtain T by deleting n in S. All n-chords in T are in order, their end points include numbers $[1, n-1]$. We call such T of the first kind if 0 is located between two n-chords, and otherwise, we call such T of the second kind. We shall show that each first kind of beautiful permutation of $[0, n-1]$ corresponds to exactly one beautiful permutation of $[0, n]$. And each second kind of permutation of $[0, n-1]$ corresponds to exactly two beautiful permutations of $[0, n]$.

If T is of the first kind, suppose that 0 is located on the arc between chords A and B. Since chords A, $(0, n)$, and B in S are in order, n must be located on the other arc between chords A and B. Thus, we can retrieve S from T uniquely. On the other hand, for each T of the first kind, we can add n in the above manner to obtain S. We shall show that the cyclic permutation S must be beautiful.

For $0 < k < n$, the k-chord of S is also the k-chord of T, so k-chords are in order.

If T is of the second kind, then the position of n to the corresponding S has two possibilities, that is, n is adjacent to 0 on either side. Similarly, we can check that S is a beautiful permutation of $[0, n]$.

Denote the total number of beautiful permutations of $[0, n]$ by M_n and the total number of the second kind of beautiful permutations of $[0, n-1]$ by L_{n-1}. Then

$$M_n = (M_{n-1} - L_{n-1}) + 2L_{n-1} = M_{n-1} + L_{n-1}.$$

It suffices to show that L_{n-1} is the number of positive integer pairs (x, y) with the constraints $x + y = n$ and $\gcd(x, y) = 1$.

For $n \geq 3$, define $\varphi(n) = |\{x : 1 \leq x \leq n, \gcd(x, n) = 1\}|$.

We shall show that $L_{n-1} = \varphi(n)$. Consider a second kind of beautiful permutation of $[0, n-1]$, marked positions $0, 1, \ldots, n-1$ clockwise on the circle such that number 0 coincides with the position 0. Denote the number of position i as $f(i)$. Suppose that $f(a) = n - 1$.

Since each number of $\{1, \ldots, n-1\}$ is at an end point of n-chord. and all n-chords are in order, by the definition of the second kind, $(1, n-1)$ is an n-chord. Thus, all n-chords are parallel. That is, $f(i) + f(n-i) = n$ for $i = 1, 2, \ldots, n-1$.

Similarly, since all $(n-1)$-chords are in order and each point is an end point of an $(n-1)$-chord, all $(n-1)$-chords are parallel. Thus, $f(i) + f(\text{mod}(a - i, n)) = n - 1$ for $i = 1, 2, \ldots, n-1$.

Therefore, $f(-i) = f(\mathrm{mod}(a - i, n)) + 1$. Consequently, by taking $i = a, 2a, \ldots, (n-1)a$ successively, and $f(0) = 0$. We have

$$f(-ak) = \mathrm{mod}(k, n), \quad k = 0, 1, 2, \ldots, (n-1). \tag{1}$$

Since f is a permutation, $\gcd(a, n) = 1$. That is, $L_{n-1} \le \varphi(n)$. To show that the equality holds for (1), we need only to show that the permutation f given by (1) is beautiful. To see this, we consider four numbers w, x, y, and z on the circle satisfying $w + y = x + z$. Their corresponding positions on the circle satisfy $(-aw) + (-ay) \equiv (-ax) + (-az)(\mathrm{mod}\,n)$, that is, the chord (w, y) and chord (x, z) are parallel. Thus, the permutation (1) is beautiful, and is of the second kind by construction.

【Score Situation】 This particular problem saw the following distribution of scores among contestants: 7 contestants scored 7 points, 4 contestants scored 6 points, 6 contestants scored 5 points, 2 contestants scored 4 points, 6 contestants scored 3 points, 6 contestants scored 2 points, 15 contestants scored 1 point, and 481 contestants scored 0 point. The average score of this problem is 0.296, indicating that it it was extremely difficult.

Among the top five teams in the team scores, the scores of this problem are as follows: the China team scored 15 points (with a total team score of 208 points), the South Korea team scored 20 points (with a total team score of 204 points), the United States team scored 17 points (with a total team score of 190 points), the Russia team scored 13 points (with a total team score of 187 points), and the North Korea team scored 11 points (with a total team score of 184 points).

The gold medal cutoff for this IMO was set at 31 points (with 45 contestants earning gold medals), the silver medal cutoff was 24 points (with 92 contestants earning silver medals), and the bronze medal cutoff was 15 points (with 141 contestants earning bronze medals).

In this IMO, no contestant achieved a perfect score of 42 points.

1.3 Summary

In the first 64 IMOs, there were a total of 19 enumerative combinatorics problems, as depicted in Figure 1.5. The score details for these problems are presented in Table 1.2. Due to the smaller number of participating teams and missing contestant score information in early IMOs, there are several blanks in Table 1.2.

These 19 problems were proposed by 16 countries. Germany proposed three problems, while Russia contributed two problems.

From Table 1.2, it can be observed that in the first 64 IMOs, there were four enumerative combinatorics problems with an average score of 0–1 point; one problem with an average score of 1–2 points; three problems with

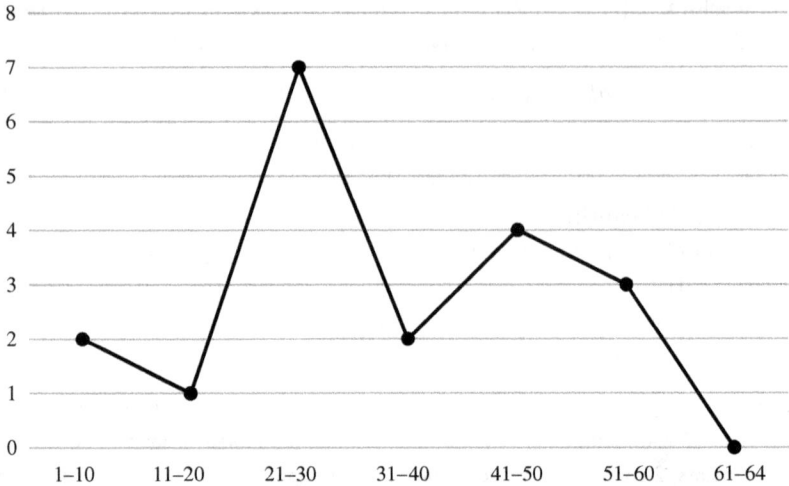

Figure 1.5 Numbers of Enumerative Combinatorics Problems in the First 64 IMOs

an average score of 2–3 points; six problems with an average score of 3–4 points; five problems with an average score above 4 points. Among these problems, the one with the lowest average score is Problem 1.16 (IMO 50-6), proposed by Russia. Overall, the enumerative combinatorics problems were relatively simple, and the contestants scored relatively high.

In the 24th–64th IMOs, there were a total of 14 enumerative combinatorics problems. Among these, three had an average score of 0–1 point; one had an average score of 1–2 points; three had an average score of 2–3 points; five had an average score of 3–4 points; two had an average score above 4 points. Further analysis of the problem numbers of these 14 enumerative combinatorics problems, as shown in Table 1.3, reveals that these problems frequently appeared as the 1st/4th problem.

From Table 1.2, it is observable that in the enumerative combinatorics problems, the average score of the top five teams generally exceeds the average score of the problem by 3 points, the average score of the 6th–15th teams typically surpasses the average score by 2 points, and the average score of the 16th–25th teams usually exceeds the average score by 1 point.

However, in early IMOs, the average score of the 16th–25th teams and the average score were very close, such as Problem 1.6 (IMO 28-1), Problem 1.7 (IMO 29-2), Problem 1.8 (IMO 30-1), Problem 1.10 (IMO 30-6), and Problem 1.11 (IMO 33-5). This phenomenon is due to the smaller number of participating teams in early IMOs. It was not until the 30th IMO in

Table 1.2 Score Details of Enumerative Combinatorics Problems in the First 64 IMOs

Problem	1.1	1.2	1.3	1.4	1.5	1.6	1.7	1.8
Full points	6	8	8	7	7	7	7	7
Average score	5.333	5.054	0.500	3.747	6.275	3.464	3.228	3.038
Top five mean			3.175	6.250	6.550	7.000	6.333	6.100
6th–15th mean			0.343	4.013	5.726	5.183	4.525	4.667
16th–25th mean					2.954	3.533	3.724	2.950
Problem number in the IMO	8-1	9-6	13-6	21-6	22-2	28-1	29-2	30-1
Proposing country	The Soviet Union	Hungary	Sweden	Germany	Germany	Germany	Czechoslovakia	The Philippines

Problem	1.9	1.10	1.11	1.12	1.13	1.14	1.15	1.16
Full points	7	7	7	7	7	7	7	7
Average score	1.337	2.526	0.757	2.735	3.449	3.558	2.077	0.168
Top five mean	4.400	5.800	3.033	6.067	6.167	6.133	6.367	1.467
6th–15th mean	2.283	3.900	1.150	5.367	5.800	5.783	4.983	0.500
16th–25th mean	0.700	2.950	0.750	3.667	5.483	4.831	3.917	0.150
Problem number in the IMO	30-3	30-6	33-5	39-2	43-1	44-1	49-5	50-6
Proposing country	Denmark	Poland	Italy	India	Colombia	Brazil	France	Russia

Problem	1.17	1.18	1.19
Full points	7	7	7
Average score	5.348	4.069	0.296
Top five mean	6.900	6.933	2.533
6th–15th mean	6.742	6.348	1.067
16th–25th mean	6.621	6.273	0.133
Problem number in the IMO	52-1	52-4	54-6
Proposing country	Mexico	Iran	Russia

Note. Top five mean = Total score of the top five teams ÷ Total number of contestants from the top five teams,
6th–15th mean = Total score of the 6th–15th teams ÷ Total number of contestants from the 6th–15th teams,
16th–25th mean = Total score of the 16th–25th teams ÷ Total number of contestants from the 16th–25th teams.

Table 1.3 Numbers of Enumerative Combinatorics Problems in the 24th–64th IMOs

Enumerative Combinatorics Problem	Problem Number			Number of Problems in the First 64 IMOs
	1, 4	2, 5	3, 6	
Enumerative combinatorics problems	6	4	4	19

1989 that the number of participating teams exceeded 50. Therefore, it is common to see situations where the average score is close to or even higher than the average score of the 16th–25th teams during this period.

Chapter 2

Existence Problems

Typically, existence problems require proving the existence or non-existence of a structure with certain specific properties. To address these problems, one can utilize theorems such as the Pigeonhole Principle, Extremal Value Principle, or Intermediate Value Theorem, or methods such as counting, coloring, mathematical induction, and proof by contradiction.

It would be ideal if the problem solver could directly construct an example; though often, such examples are not easy to construct. Similarly, in fields like number theory and algebra, many existence problems can be found, and the approaches to solving them are similar.

In the first 64 IMOs, there had been a total of 13 existence problems, accounting for approximately 14.9% of all combinatorics problems. The statistical distribution of these problems in the previous IMOs is presented in Table 2.1.

Due to the fundamental nature, the existence problems appeared frequently in early IMOs. And many existence problems are also combinatorial geometry problems, which are not included in this chapter.

It is important to note that for each problem, the solutions are followed by information on the scores, including the number of contestants in each score range, the average score, and the scores of the top five teams. However, early IMOs often lacked information on contestant scores, so the number of contestants in each score range only represents the counted number of contestants, and some problems lack scores of the top five teams.

Table 2.1 Numbers of Existence Problems in the First 64 IMOs

Content	Session							Total
	1–10	11–20	21–30	31–40	41–50	51–60	61–64	
Existence problems	0	3	2	1	3	3	1	13
Combinatorics problems	7	11	16	16	11	19	7	87
Percentage of existence problems among combinatorics problems	0.0%	27.3%	12.5%	6.3%	27.3%	15.8%	14.3%	14.9%

2.1 Common Theorems, Formulas, and Methods

2.1.1 *Common theorems*

(1) *Extremal principle*

(i) Let $\mathbf{N}_+$ be the set of all positive integers and M be a non-empty subset of $\mathbf{N}_+$ (M may be finite or infinite). Then M has a minimum element.

(ii) Let T be a non-empty finite set of real numbers. Then T has a minimum element and a maximum element.

(2) *Pigeonhole principle*

(i) If m items are placed in n boxes, then there is a box containing at least $\left[\frac{m-1}{n}\right] + 1$ items.

(ii) If m items are placed in n boxes, then there is a box containing at most $\left[\frac{m}{n}\right]$ items.

(3) *Average principle*

(i) Let $a_1, a_2, \ldots, a_n$ be real numbers with the arithmetic mean $A = \frac{1}{n}(a_1 + a_2 + \cdots + a_n)$. Then there exists one number of $a_1, a_2, \ldots, a_n$ which is $\leq A$ (also a number $\geq A$).

(ii) Let $a_1, a_2, \ldots, a_n$ be positive reals with the geometric mean $G = \sqrt[n]{a_1 a_2 \cdots a_n}$. Then there exists one number of $a_1, a_2, \ldots, a_n$ which is $\leq G$ (also a number $\geq G$).

(4) *Overlapping principle*

Place n two-dimensional regions $A_1, A_2, \ldots, A_n$ with areas $S_1, S_2, \ldots, S_n$ into a region A with area S.

(i) If $S_1 + S_2 + \cdots + S_n > S$, then there are two regions A_i and $A_j (1 \leq i < j \leq n)$ that share a common point.

(ii) If $S_1 + S_2 + \cdots + S_n < S$, then there exists a point in A that does not belong to any of $A_1, A_2, \ldots, A_n$.

(5) *Intermediate value principle*

(i) **Continuous version:** Let $f(x)$ be a continuous function on the closed interval $[a, b]$. If $f(a) < A < f(b)$ (or $f(a) > A > f(b)$), then there exists $c \in (a, b)$ such that $f(c) = A$.

(ii) **Discrete version:** Let $a_1, a_2, \ldots, a_n$ be a sequence of integers, and A and B be two integers satisfying the following conditions:

(1) $a_1 < A < B < a_n$ (resp., $a_1 > A > B > a_n$);

(2) for every $i (1 \leq i \leq n - 1)$, there holds $a_{i+1} - a_i \leq 1$ (resp., $a_{i+1} - a_i \geq -1$), then for every integer C with $A \leq C \leq B$ (resp., $A \geq C \geq B$), there exists some $i_0 (1 < i_0 < n)$ such that $a_{i_0} = C$.

Proof. We only consider the case $a_1 < A \leq C \leq B < a_n$; the other case is similar. Assume on the contrary that $a_i \neq C$ for every $1 < i < n$. Since $a_1 < C < a_n$, there exists $k (1 \leq k < n)$ such that $a_k \leq C - 1$ and $a_{k+1} \geq C + 1$. Hence, $a_{k+1} - a_k \geq 2$, a contradiction. This completes the proof.

2.1.2 *Common methods*

(1) *Applying extremal principle*

Example 2.1. Given 100 points in a plane. It is known that the distance between any two points is not larger than 1 and the triangle formed by any three points is obtuse-angled. Show that the points can be covered by a closed disc of radius $\frac{1}{2}$.

Proof. Choose two points with a maximum distance, since there are finitely many choices. Let A and B be two points with the maximum distance. Consider the circle with diameter AB. Let R be its radius. By assumption, $AB \leq 1$. Hence $R \leq \frac{1}{2}$.

For any C other than A and B, since $AB \geq AC$, $AB \geq BC$, and $\triangle ABC$ is an obtuse-angled triangle, $\angle C$ must be obtuse. Hence C lies in the circle with diameter AB. Thus the 100 points are covered by this disc.

(2) *Applying pigeonhole principle*

Example 2.2 (Ramsey's Theorem). Given 6 points in the space such that no 4 points are coplanar. Each segment between 2 of these 6 points is colored red or blue. Then there exists a triangle whose edges are of the same color.

Proof. Consider the five segments with the common endpoint A; they are colored in two colors. By the pigeonhole principle, three of them are of the same color, say that AB, AC, and AD are red. Consider the sides of $\triangle BCD$. If one of them is red, say BD, then $\triangle ABD$ is a red triangle. Otherwise, $\triangle BCD$ is a blue triangle.

(3) *Applying intermediate value principle*

Example 2.3. Let F be a planar convex figure. There exists a line that bisects both its area and perimeter.

Proof. Choose a directed line l_0 (a line l_0 with a fixed positive direction). We first claim that there exists a line, parallel to l_0, which bisects the perimeter of F. Indeed, F is bounded by two lines l_1 and l_2 parallel to l_0. Let d be the distance between l_1 and l_2, and let l_x be a parallel line between l_1 and l_2 such that the distance between l_x and l_1 is x. Let $f(x)$ be the difference of the boundary lengths of F lying between l_x and l_1, and between l_x and l_2. Thus $f(x)$ is continuous on $[0, d]$, $f(0) < 0$, and $f(d) > 0$. By the intermediate value principle, there exists $x_0 \in (0, d)$ such that $f(x_0) = 0$, that is l_{x_0} bisects the boundary of F.

Assume that l_0 bisects the boundary of F. For each direction θ (rotating from l_0 counterclockwise by angle θ), let l_θ be the line with direction θ which bisects the boundary of F. For $\theta \in [0, \pi]$, let $g(\theta)$ be the difference of areas of F lying above and below the line l_θ.

Since $g(\theta)$ is continuous on $[0, \pi]$, and $g(\pi) = -g(0)$, it follows from the intermediate value theorem again that there is some $\theta_0 \in [0, \pi]$ such that $g(\theta_0) = 0$. That is l_{θ_0} bisects the area of F and the perimeter as well.

(4) *Proof by contradiction*

Example 2.4. Nine mathematicians met at an international mathematics conference and found that at least two out of any three of them could converse in the same language. If each mathematician can speak at most three languages, prove that at least three of these mathematicians can converse in the same language.

Proof. Assume that there are no three people who can speak the same language. Then each language can only be spoken by at most two people.

Let the nine mathematicians be $A_1, A_2, \ldots, A_9$. Since A_1 speaks at most three languages, A_1 can converse with at most three people. Assume that A_1 cannot converse with A_2, A_3, A_4, A_5, and A_6. Since A_2 can speak at most three languages, he cannot converse with at least one of A_3, A_4, A_5, and A_6, say A_3. Then no two of A_1, A_2, and A_3 can converse in the same language. Contrary to the assumption.

Thus, the assertion is true.

(5) *Mathematical induction*

Example 2.5. Color each vertex of a convex 2003 polygon with a color, such that any two adjacent vertices are of different colors. Prove that the polygon can be divided into triangles using diagonals that do not intersect with each other, so that the two ends of each diagonal are of different colors.

Proof. Consider the general problem concerning a polygon with $n = 2k+1$ ($k \in \mathbf{N}_+$) vertices. We prove the conclusion by induction on k.

For $k = 1$, we have $n = 3$. Then it is obvious.

Assume the assertion is true for $k = m$. Consider the case $k = m + 1$. Suppose the vertices of a convex $(2m + 3)$-gon is colored in some way. There exists a vertex A whose two neighboring vertices are of different colors. For otherwise if two neighboring vertices are of the same color, then the color of vertices alternates, which is impossible since n is odd. Draw the diagonal connecting the two neighboring vertices of A and dissect the polygon into a triangle and a $(2m + 2)$-gon.

If there is some vertex B of the $(2m + 2)$-gon whose two neighboring vertices are of different colors, then connect these two vertices and dissect the $(2m + 2)$-gon into a triangle and a $(2m + 1)$-gon, and we can apply the inductive hypothesis to the $(2m + 1)$-gon. Otherwise, the colors on the $(2m + 2)$-gon alternate. And vertex A is of a different color. We can draw all diagonals from A to dissect the original polygon into triangles satisfying the requirement.

(6) *Constructive proof*

Example 2.6. Is it possible to have a figure as shown in Figure 2.1 such that the sides of four triangles and a pentagon (total 13 segments) have distinct integer lengths?

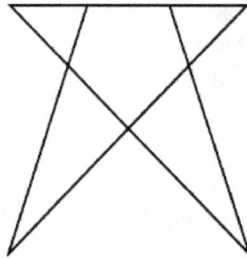

Figure 2.1   A 4-Star Figure

Solution. The answer is in the affirmative. We construct an example such that the four triangles are all right-angled triangle with sides 3:4:5. Let a, b, and c be distinct integers greater than 1 and m be a positive integer, as shown in Figure 2.2.

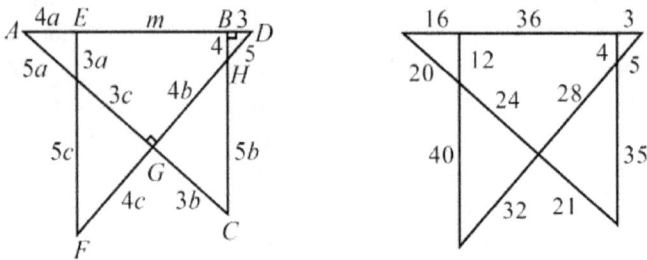

Figure 2.2   An Example of 4-Star Figure

Since $\triangle ABC \backsim \triangle AGD \backsim \triangle FED \backsim \triangle HBD$,

$$(4 + 5b) : (4a + m) : (5a + 3c + 3b) = 3 : 4 : 5,$$

$$(5 + 4b) : (5a + 3c) : (3 + m + 4a) = 3 : 4 : 5,$$

$$(3 + m) : (3a + 5c) : (5 + 4b + 4c) = 3 : 4 : 5.$$

It follows from the three equalities above that $4 + 5b$ and $5 + 4b$ are divisible by 3. So, we may assume $b = 3k + 1(k \in \mathbf{N}_+)$. Then $4a + m = 20k + 12$ and $5a + 3c = 16k + 12$.

Take $a = 2k$, $m = 12k + 12$, and $c = 2k + 4$. It is easy to verify that the three equalities above hold. If we take $k = 2$, then the 13 segments are of different lengths, and we obtain a figure with the prescribed property.

(7) *Counting method*

Example 2.7. Given 9 points in a plane with no three collinear points. Connect 21 segments with endpoints being the given points. Show that there exists a triangle whose sides are among these 21 segments.

Proof. Assume on the contrary that the assertion is false. Assume that the number of segments with endpoint A_1 is a maximum, say there are k such segments, which are $A_1B_1, A_1B_2, \ldots, A_1B_k$. The remaining points are denoted $A_2, A_3, \ldots, A_{9-k}$.

Since there is no triangle, $B_1, B_2, \ldots, B_k$ are pairwise non-adjacent. Hence the number of segments with endpoints B_i is at most $9 - k$ (since B_i can only be adjacent to some of $A_1, A_2, \ldots, A_{9-k}$). The number of segments emanating from each of $A_1, A_2, \ldots, A_{9-k}$ is at most k. Thus, the number of segments is at most $\frac{k(9-k)+(9-k)k}{2} = k(9 - k) \leq 20$. Contrary to the fact that there are 21 segments. Thus, the assertion is true.

2.2 Problems and Solutions

Problem 2.1 (IMO 12-4, proposed by Czechoslovakia). Find the set of all positive integers n with the property that the set $\{n, n + 1, n + 2, n + 3, n + 4, n + 5\}$ can be partitioned into two sets such that the product of the numbers in one set equals the product of the numbers in the other set.

Answer. The required n does not exist.

Solution 1. A set of six consecutive integers contains at least one number divisible by 5. Since the set can be partitioned into two sets with equal products, it contains two numbers divisible by 5, and these two numbers must be n and $n + 5$. Note that n and $n + 5$ are separated, and $n(n + 1) > n + 5$ for $n \geq 5$, the set must be partitioned into two sets with equal size. It is left with the following two cases; other cases can be ruled out quite easily.

(a) $\{n, n+2, n+4\} \cup \{n+1, n+3, n+5\}$.
(b) $\{n, n+3, n+4\} \cup \{n+1, n+2, n+5\}$.

For case (a), the two sets cannot have equal products since $n < n+1$, $n+2 < n+3$, and $n+4 < n+5$.

For case (b), we have

$$n(n+3)(n+4) = (n+1)(n+2)(n+5).$$

After simplification, we get $n^2 + 5n + 10 = 0$. However, this equation has no real roots.

Solution 2. Assume the contrary that such n exists. It follows from the assumption that if a prime divides one number of

$$n, \quad n+1, \quad n+2, \quad n+3, \quad n+4, \text{ and } n+5,$$

then it divides another number as well. Consequently, none of these numbers is divisible by a prime greater than 5.

As in the Solution 1, we know that n and $n + 5$ are divisible by 5. Hence the numbers $n + 1$, $n + 2$, $n + 3$, and $n + 4$ can only have prime divisors 2 and 3. Since two of these four consecutive integers are odd, they must be powers of 3, which is impossible since the difference of two powers of 3 cannot be 2.

This verifies our claim.

【Score Situation】This particular problem saw the following distribution of scores among contestants: 15 contestants scored 6 points, 11 contestants scored 5 points, 5 contestants scored 4 points, 1 contestant scored 3 points, 2 contestants scored 2 points, 1 contestant scored 1 point, and no contestant scored 0 point. The average score of this problem is 4.943, indicating that it was simple.

Among the top five teams in the team scores, the Hungary team achieved a total score of 233 points, the German Democratic Republic team achieved a total score of 221 points, the Soviet Union team achieved a total score of 221 points, the Yugoslavia team achieved a total score of 209 points, and the Romania team achieved a total score of 208 points.

The gold medal cutoff for this IMO was set at 37 points (with 7 contestants earning gold medals), the silver medal cutoff was 30 points (with 11 contestants earning silver medals), and the bronze medal cutoff was 19 points (with 40 contestants earning bronze medals).

In this IMO, only three contestants achieved a perfect score of 40 points, namely Wolfgang Burmeister from the German Democratic Republic, Imre Ruzsa from Hungary, and Andrei Hodulev from the Soviet Union.

Problem 2.2 (IMO 14-1, proposed by the Soviet Union). Prove that from a set of ten distinct two-digit numbers (in the decimal system), it is possible to select two disjoint subsets whose members have the same sum.

Proof. This set has $2^{10} - 1 = 1023$ nonempty subsets. The sum of elements of each subset is not greater than

$$99 + 98 + \cdots + 90 = 945.$$

There exist two subsets with equal sum of elements by the pigeonhole principle. By removing the common numbers of these two subsets if necessary, we obtain two disjoint subsets with equal sums of their members.

【Score Situation】This particular problem saw the following distribution of scores among contestants: 19 contestants scored 5 points, no contestant scored 4 points, 1 contestant scored 3 points, no contestant scored 2 points, 3 contestants scored 1 point, and 10 contestants scored 0 point. The average score of this problem is 3.061, indicating that it was relatively straightforward.

Among the top five teams in the team scores, the scores of this problem are as follows: the Soviet Union team scored 36 points (with a total team score of 270 points), the Hungary team scored 35 points (with a total team score of 263 points), the German Democratic Republic team scored 40 points (with a total team score of 239 points), the Romania team scored 16 points (with a total team score of 208 points), and the United Kingdom team scored 35 points (with a total team score of 179 points).

The gold medal cutoff for this IMO was set at 40 points (with 8 contestants earning gold medals), the silver medal cutoff was 30 points (with 16 contestants earning silver medals), and the bronze medal cutoff was 19 points (with 30 contestants earning bronze medals).

In this IMO, a total of eight contestants achieved a perfect score of 40 points.

Problem 2.3 (IMO 20-6, proposed by the Netherlands). An international society has its members from six different countries. The list of members contains 1978 names, numbered $1, 2, \ldots, 1978$. Prove that there is at least one member whose number is the sum of the numbers of two members from his own country, or twice as large as the number of one member from his own country.

Proof. The problem may be stated as follows: if the set $\{1, 2, 3, \ldots, 1978\}$ is separated in any way into six disjoint subsets A, B, C, D, E, and F, then at least one of these six sets contains the sum of two (possibly equal) elements of that set.

First, we observe that no matter how 1978 elements are distributed among six sets, at least one of these sets, say A, contains at least $\lceil \frac{1978}{6} \rceil = 330$ members, which we denote by $a_1 < a_2 < \cdots < a_{330}$. If any of the 329 differences

$$b_1 = a_{330} - a_{329}, b_2 = a_{330} - a_{328}, \ldots, b_{329} = a_{330} - a_1,$$

belongs to A, then the problem is solved, because for some k,

$$a_{330} - a_k = a_l \in A, \text{ so } a_k + a_l = a_{330}.$$

So, suppose none of the 329 differences $a_{330} - a_i$ belongs to A; instead, they are distributed among the five remaining sets. Again, we conclude that at least one of these, say B, contains at least $\lceil \frac{329}{5} \rceil = 66$ of these differences. We take their differences

$$c_j = b'_{66} - b'_{66-j}.$$

We describe this process in more detail: We order the differences $b_j = a_{330} - a_{330-j}$ which are in B according to increasing size and rename them $b'_1, b'_2, \ldots, b'_{66}$. We form

$$c_j = b'_{66} - b'_{66-j} = (a_{330} - a_i) - (a_{330} - a_k) = a_k - a_i,$$

$$j = 1, 2, \ldots, 65; \quad i < k < 330; \quad 1 \le a_i < a_k < 1978.$$

Thus the c_j are integers satisfying $1 \le c_j < 1977$.

If any of these c_j occur in either A or B, then the problem is solved. If not, they are distributed among the four remaining sets at least one of which, say C, contains at least $\lceil \frac{65}{4} \rceil = 17$ of them. Again, we order their 16 differences $c_{17} - c_j$ according to the size and rename them c'_i so that $c'_i < c'_j$ for $i < j$. If none are elements of sets A, B, or C, then all are distributed among the three remaining sets, at least one of which, say D, contains at least $\lceil \frac{17}{3} \rceil = 6$ elements, called d'_i.

Similarly, we argue that if their five differences are not in A, B, C, or D, then at least one of the two remaining sets, say E, contains at least $\lceil \frac{5}{2} \rceil = 3$ of them, and if neither of their two differences e'_1 and e'_2 is in one of the sets A, B, C, D, and E, then they must both be in F. Their difference $e'_2 - e'_1 < 1978$ must belong to one of the six sets.

【Score Situation】 This particular problem saw the following distribution of scores among contestants: 2 contestants scored 8 points, no contestant scored 7 points, no contestant scored 6 points, 0 contestant scored 5 points, 2 contestants scored 4 points, 2 contestants scored 3 points, 1 contestant scored 2 points, 12 contestants scored 1 point, and 29 contestants scored 0 point. The average score of this problem is 0.917, indicating that it was extremely difficult.

Among the top five teams in the team scores, the scores of this problem are as follows: the Romania team scored 36 points (with a total team score of 237 points), the United States team scored 11 points (with a total team score of 225 points), the United Kingdom team scored 12 points (with a total team score of 201 points), the Vietnam team scored 12 points (with a total team score of 200 points), and the Czechoslovakia team scored 8 points (with a total team score of 195 points).

The gold medal cutoff for this IMO was set at 35 points (with 5 contestants earning gold medals), the silver medal cutoff was 27 points (with 20 contestants earning silver medals), and the bronze medal cutoff was 22 points (with 38 contestants earning bronze medals).

In this IMO, only one contestant achieved a perfect score of 40 points, namely Mark Kleiman from the United States.

Problem 2.4 (IMO 24-5, proposed by Poland). Is it possible to choose 1983 distinct positive integers, all less than or equal to 10^5, no three of which are consecutive terms of an arithmetic progression? Justify your answer.

Proof 1. We construct a set T with more than 1983 positive integers, all less than 10^5 such that no three are in an arithmetic progression, that is, no three satisfy $x + z = 2y$.

The set T consists of all positive integers whose base 3 representations have at most 11 digits, each of which is either 0 or 1 (i.e., no 2's). There are $2^{11} - 1 > 1983$ of them, and the largest is

$$1 + 3^2 + 3^3 + \cdots + 3^{10} = 88573 < 10^5.$$

Now suppose $x + z = 2y$ for some $x, y, z \in T$. The number $2y$ for any $y \in T$ consists only of the digits 0 and 2. Hence x and z must match digit for digit, and it follows that $x = z = y$. Hence T contains no arithmetic progressions of length 3, and the desired selection is possible.

Proof 2. It is easy to show that if $a_1 < a_2 < \cdots < a_k$ are integers in $\{1, 2, \ldots, n\}$ such that no three form an arithmetic progression, then the following $2k$ numbers in $\{1, 2, \ldots, 3n\}$ share the same property:

$$a_1 < a_2 < \cdots < a_k < 2n + a_1 < 2n + a_2 < \cdots < 2n + a_k.$$

Since 1, 2, 3, and 4 from $\{1, 2, 3, 4, 5\}$ contain no arithmetic progression of length 3, we can find 4×2^k numbers among the first 5×3^k positive integers with no arithmetic progression of length 3. Put $k = 9$, and noting that

$$5 \times 3^9 = 98415 < 10^5, \quad 4 \times 2^9 = 2408 > 1983,$$

we conclude that the desired selection is possible.

【Score Situation】 This particular problem saw the following distribution of scores among contestants: 10 contestants scored 7 points, 1 contestant scored 6 points, no contestant scored 5 points, 1 contestant scored 4 points, 1 contestant scored 3 points, 1 contestant scored 2 points, 1 contestant scored 1 point, and 25 contestants scored 0 point. The average score of this problem is 2.150, indicating that it had a certain level of difficulty.

Among the top five teams in the team scores, the Germany team achieved a total score of 212 points, the United States team achieved a total score of 171 points, the Hungary team achieved a total score of 170 points, the Soviet Union team achieved a total score of 169 points, and the Romania team achieved a total score of 161 points.

The gold medal cutoff for this IMO was set at 38 points (with 9 contestants earning gold medals), the silver medal cutoff was 26 points (with 27 contestants earning silver medals), and the bronze medal cutoff was 15 points (with 57 contestants earning bronze medals).

In this IMO, a total of four contestants achieved a perfect score of 42 points.

Problem 2.5 (IMO 26-4, proposed by Mongolia). Given a set M of 1985 distinct positive integers, none of which has a prime divisor greater than 26. Prove that M contains at least one subset of four distinct elements whose product is the fourth power of an integer.

Proof. Since there are only 9 primes less than 26, each of the 1985 members of M has a prime factorization in which at most 9 distinct primes occur:

$$m = p_1^{k_1} p_2^{k_2} \cdots p_9^{k_9}, \tag{1}$$

where k_i are non-negative integers. For each element of M we assign the 9-tuple $(x_1, x_2, \ldots, x_9)$, where $x_i = 0$ or 1 depending on whether k_i is even or odd.

Thus 2^9 distinct 9-tuples are possible. By the pigeonhole principle, any subset of $2^9 + 1$ elements of M contains at least two distinct integers, say a_1 and b_1, assigned with the same 9-tuple. It follows that their product is a perfect square: $a_1 b_1 = c_1^2$.

When we remove such a pair a_1, b_1 from the set M, we are left with $1985 - 2 = 1983 > 2^9 + 1$ numbers. Apply the pigeonhole principle again and continue removing such pairs as long as at least $2^9 + 1$ numbers are left in M. Since $1985 > 3 \times (2^9 + 1)$, we can remove $2^9 + 1$ pairs a_i, b_i and still have at least $2^9 + 1$ numbers left in M.

We now look at the $2^9 + 1$ removed pairs (a_i, b_i), and take the square roots $c_i = \sqrt{a_i b_i}$ with $1 \leq i \leq 2^9 + 1$. The c_i cannot contain prime factors other than $p_1, p_2, \ldots, p_9$, so there is at least one pair c_i and c_j whose product is a square, say $c_i c_j = d^2$. It follows that

$$a_i b_i a_j b_j = c_i^2 c_j^2 = d^4$$

for some a_i, a_j, b_i, and b_j in M.

Remark. The same argument yields the more general result: if the prime divisors of M are limited to n distinct primes, and if M contains at least $3(2^n + 1)$ distinct elements, then M contains a subset of four elements whose product is the fourth power of some integer.

【Score Situation】 This particular problem saw the following distribution of scores among contestants: 32 contestants scored 7 points, 2 contestants scored 6 points, 6 contestants scored 5 points, 16 contestants scored 4 points, 18 contestants scored 3 points, 46 contestants scored 2 points, 28 contestants scored 1 point, and 61 contestants scored 0 point. The average score of this problem is 2.411, indicating that it had a certain level of difficulty.

Among the top five teams in the team scores, the scores of this problem are as follows: the Romania team scored 35 points (with a total team score of 201 points), the United States team scored 30 points (with a total team score of 180 points), the Hungary team scored 30 points (with a total team score of 168 points), the Bulgaria team scored 22 points (with a total team score of 165 points), and the Vietnam team scored 25 points (with a total team score of 144 points).

The gold medal cutoff for this IMO was set at 34 points (with 14 contestants earning gold medals), the silver medal cutoff was 22 points (with 35 contestants earning silver medals), and the bronze medal cutoff was 15 points (with 52 contestants earning bronze medals).

In this IMO, only two contestants achieved a perfect score of 42 points, namely Géza Kós from Hungary and Daniel Tătaru from Romania.

Problem 2.6 (IMO 38-4, proposed by Iran). An $n \times n$ matrix whose entries come from the set $S = \{1, 2, \ldots, 2n - 1\}$ is called a *silver* matrix if, for each $i = 1, 2, \ldots, n$, the ith row and the ith column together contain all elements of S. Show that:

(a) there is no silver matrix for $n = 1997$;
(b) silver matrices exist for infinitely many values of n.

Proof. (1) Let A be an $n \times n$ silver matrix with $n > 1$. Since all $2n - 1$ numbers of S appear in A, and there are only n numbers on the diagonal, we choose and fix an element $x \in S$ that does not appear on the main diagonal of A. For each $i = 1, 2, \ldots, n$, let A_i be the set of all entries of A lying on the ith row or ith column, which we call the ith "cross."

Each cross A_i contains exactly one x. If some x is on the ith row and jth column ($i \neq j$), then this x belongs to A_i and A_j. We may pair up

A_i and A_j in this way. Thus all n "crosses" of A are matched into pairs. Consequently, n is even. However, 1997 is odd, so there is no silver matrix for $n = 1997$.

(2) For $n = 2$,

$$A = \begin{pmatrix} 1 & 2 \\ 3 & 1 \end{pmatrix}$$

is a silver matrix. For $n = 4$,

$$A = \begin{pmatrix} 1 & 2 & 5 & 6 \\ 3 & 1 & 7 & 5 \\ 4 & 6 & 1 & 2 \\ 7 & 4 & 3 & 1 \end{pmatrix}$$

is a silver matrix.

Suppose that there exists an $n \times n$ silver matrix A such that the main diagonal consists of only 1. We construct a $2n \times 2n$ silver matrix D as follows:

$$D = \begin{pmatrix} A & B \\ C & A \end{pmatrix},$$

where B is an $n \times n$ matrix obtained from A by adding $2n$ to each entry and C is obtained from B by replacing all entries on the main diagonal with $2n$.

Consider the ith "cross" of D. Assume $i \leq n$ and it is similar for $i > n$. The ith "cross" of D consists of the ith "cross" of A, the ith row of B, and the ith column of C. The ith "cross" of A contains all the numbers of $\{1, 2, \ldots, 2n - 1\}$, while the ith row of B and the ith column of C contain all the numbers of $\{2n, 2n + 1, \ldots, 4n - 1\}$.

Thus D is a $2n \times 2n$ silver matrix with 1 on the main diagonal. Therefore an $n \times n$ silver matrix exists for every $n = 2^k$.

【Score Situation】 This particular problem saw the following distribution of scores among contestants: 135 contestants scored 7 points, 31 contestants scored 6 points, 29 contestants scored 5 points, 58 contestants scored 4 points, 34 contestants scored 3 points, 20 contestants scored 2 points, 72 contestants scored 1 point, and 81 contestants scored 0 point. The average score of this problem is 3.743, indicating that it was relatively straightforward.

Among the top five teams in the team scores, the scores of this problem are as follows: the China team scored 38 points (with a total team score of 223 points), the Hungary team scored 42 points (with a total team score of 219 points), the Iran team scored 42 points (with a total team score of 217 points), the United States team scored 42 points (with a total team score of 202 points), and the Russia team scored 30 points (with a total team score of 202 points).

The gold medal cutoff for this IMO was set at 35 points (with 39 contestants earning gold medals), the silver medal cutoff was 25 points (with 70 contestants earning silver medals), and the bronze medal cutoff was 15 points (with 122 contestants earning bronze medals).

In this IMO, a total of four contestants achieved a perfect score of 42 points.

Problem 2.7 (IMO 42-3, proposed by Germany). Twenty-one girls and 21 boys took part in a mathematical contest.

- Each contestant solved at most six problems.
- For each girl and each boy, at least one problem was solved by both.

Prove that there was a problem that was solved by at least three girls and at least three boys.

Proof 1. We construct a table of size 21×20, as shown in Figure 2.3. Fill the problem solved by both the ith girl and the jth boy (if there are more than 1 such problem, choose any one of such problems) in the (i, j) entry ($1 \le i \le 21$ and $1 \le j \le 20$).

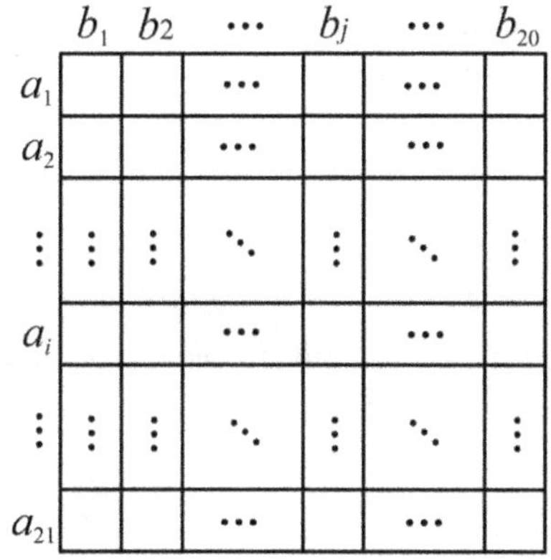

Figure 2.3 A Table of Size 21×20

Since each boy solved at most six problems, at most six distinct problems appear on each column. For each column, we color the entries on that column blue if the problem of that entry appears at least three times on that column. By the pigeonhole principle, on each column there is a problem appearing at least ($\lceil \frac{21}{6} \rceil + 1$) = 4 times, and hence there exist blue entries on each column.

If the number of blue entries is no more than 10, then there are at least 11 entries that are not blue, with at most 5 distinct problems. There is a problem appearing at least $[\frac{11}{5}]+1 = 3$ times and such entry is supposed to be blue. A contradiction. Thus, there are at least 11 blue entries on each column, and at least 220 blue entries in the table.

Since each girl solved at most six problems, at most six distinct problems appear on each row. We color an entry red if the problem in that entry appears at least three times in its row. Similarly, we conclude that there are at least 10 red entries on each row and at least 210 red entries in the table.

Since there are a total of 420 entries and $220 + 210 = 430 > 420$, there is an entry which is colored both blue and red. The problem in that entry is solved by at least 3 boys and 3 girls.

Proof 2. Suppose on the contrary that each problem is solved by at most 2 boys or by at most 2 girls.

Let $A = \{A_1, A_2, \ldots, A_k\}$ be the set of problems that are solved by at most 2 girls and $B = \{A_{k+1}, A_{k+2}, \ldots, A_{k+m}\}$ be the set of remaining problems that are solved by at most 2 boys.

We denote the 21 boys by $p_1, p_2, \ldots, p_{21}$ and the 21 girls by $q_1, q_2, \ldots, q_{21}$. Assume that there are x_j contestants who solved A_j for $1 \le j \le k + m$. By assumption,

$$x_1 + x_2 + \cdots + x_{k+m} \le 6 \times 42 = 252.$$

We may assume that each problem is solved by at least one boy and at least one girl; otherwise we can just delete that problem. Thus,

$$x_j \ge 2, \quad 1 \le j \le k + m.$$

We pair up p_i and q_j for certain problem if they solved that problem. By assumption there are 21^2 pairs. For each A_j, we have at most

$$\max\{(x_j - 1) \times 1, (x_j - 2) \times 2\} \le 2x_j - 3$$

pairs for A_j. Hence,

$$21^2 \le (2x_1 - 3) + (2x_2 - 3) + \cdots + (2x_{k+m} - 3)$$

$$= 2(x_1 + x_2 + \cdots + x_{k+m}) - 3(k + m)$$

$$\le 2 \times 6 \times 42 - 3(k + m).$$

As a result,

$$k + m \le 21. \tag{1}$$

On the other hand, each p_j solved at most six problems in A. Thus there are at most $2 \times 6 = 12$ girls paired up with p_j for some problem in A.

Therefore, at least $21 - 12 = 9$ girls are paired up with p_j for some problems in B. Hence p_j solved at least one problem in B. Since each problem of B is solved by at most 2 boys, $21 \leq 2m$, i.e., $m \geq 11$.

Similarly, $k \geq 11$. It follows that $m + k \geq 22$, which contradicts (1). The proof is completed.

【Score Situation】 This particular problem saw the following distribution of scores among contestants: 20 contestants scored 7 points, 3 contestants scored 6 points, 6 contestants scored 5 points, 1 contestant scored 4 points, 8 contestants scored 3 points, 36 contestants scored 2 points, 127 contestants scored 1 point, and 272 contestants scored 0 point. The average score of this problem is 0.877, indicating that it was extremely difficult.

Among the top five teams in the team scores, the scores of this problem are as follows: the China team scored 23 points (with a total team score of 225 points), the United States team scored 28 points (with a total team score of 196 points), the Russia team scored 25 points (with a total team score of 196 points), the Bulgaria team scored 12 points (with a total team score of 185 points), and the South Korea team scored 9 points (with a total team score of 185 points).

The gold medal cutoff for this IMO was set at 30 points (with 39 contestants earning gold medals), the silver medal cutoff was 20 points (with 81 contestants earning silver medals), and the bronze medal cutoff was 11 points (with 122 contestants earning bronze medals).

In this IMO, a total of four contestants achieved a perfect score of 42 points.

Problem 2.8 (IMO 46-6, proposed by Romania). In a mathematical competition, in which six problems were posed to the participants and every two of these problems were solved by more than $\frac{2}{5}$ of the contestants. Moreover, no contestant solved all the six problems. Show that there are at least two contestants who solved exactly five problems each.

Proof. Assume that there were n contestants $C_1, C_2, \ldots, C_n$ and the six problems were $P_1, P_2, \ldots, P_6$.

Let $S = \{(C_k; P_i, P_j) | 1 \leq k \leq n, 1 \leq i < j \leq 6, C_k$ solved both P_i and $P_j\}$. Now we will count $|S|$.

Let x_{ij} be the number of contestants having solved both P_i and $P_j (1 \leq i < j \leq 6)$. By hypothesis, $x_{ij} > \frac{2}{5}n \Leftrightarrow x_{ij} \geq \frac{2n+1}{5}$. Therefore

$$|S| = \sum_{1 \leq i < j \leq 6} x_{ij} \geq \sum_{1 \leq i < j \leq 6} \frac{2n+1}{5} = C_6^2 \cdot \frac{2n+1}{5} = 6n + 3. \quad (1)$$

If the number of contestants who solved exactly five problems was at most one, then by hypothesis, no contestant has solved all problems, so the other contestants each solved four problems at most.

Let $a_1, a_2, \ldots, a_n$ be the number of problems solved by contestants $C_1, C_2, \ldots, C_n$ respectively. Without loss of generality, we can assume that $5 \geq a_1 \geq a_2 \geq \cdots \geq a_n \geq 0$.

If $a_1 \leq 4$, then $a_k \leq 4\,(1 \leq k \leq n)$, so

$$|S| = \sum_{k=1}^{n} C_{a_k}^2 \leq \sum_{k=1}^{n} C_4^2 = 6n,$$

which is an contradiction to (1). Therefore $a_1 = 5$, $4 \geq a_2 \geq \cdots \geq a_n$.

It is trivial that $n \geq 2$. If $a_n \leq 3$, then

$$|S| = \sum_{k=1}^{n} C_{a_k}^2 \leq C_5^2 + (n-2)C_4^2 + C_3^2 = 6n + 1.$$

This is also in contradiction with (1). So $a_n \geq 4$, i.e., $a_2 = a_3 = \cdots = a_n = 4$.

Assume, without loss of generality, that the five problems C_1 solved were $P_1, P_2, \ldots, P_5$ and the number of contestants who solved the problem $P_j(1 \leq j \leq 6)$ was $b_j(1 \leq j \leq 6)$. Then

$$b_1 + b_2 + \cdots + b_6 = a_1 + a_2 + \cdots + a_n = 5 + 4(n-1) = 4n + 1. \tag{2}$$

Consider

$$\sum_{j=2}^{6} x_{1j} \geq \sum_{j=2}^{6} \frac{2n+1}{5} = 2n + 1. \tag{3}$$

Since the problem P_1 has been solved by b_1 contestants, and one of them solved five problems and the others solved four problems,

$$\sum_{j=2}^{6} x_{1j} = 4 + 3(b_1 - 1) = 3b_1 + 1. \tag{4}$$

Combining (3) and (4), we obtain $2n + 1 \leq 3b_1 + 1$, which can be rewritten as $b_1 \geq \frac{2n}{3}$.

Similarly,

$$b_k \geq \frac{2n}{3}\,(k = 1, 2, 3, 4, 5). \tag{5}$$

Consider

$$\sum_{j=1}^{5} x_{j6} \geq \sum_{j=1}^{5} \frac{2n+1}{5} = 2n + 1. \tag{6}$$

By a similar argument, since exactly b_6 contestants solved P_6 and each of these contestants solved four problems,

$$\sum_{j=1}^{5} x_{j6} = 3b_6. \tag{7}$$

Together with (6) and (7) we obtain $3b_6 \geq 2n + 1$, which we can rewrite as

$$b_6 \geq \frac{2n+1}{3}. \tag{8}$$

If $n \not\equiv 0 \pmod 3$, then $\frac{2n}{3}$ is not an integer. By (5) we obtain

$$b_k > \frac{2n}{3} (1 \leq k \leq 5),$$

which implies

$$b_k \geq \frac{2n+1}{3} (1 \leq k \leq 5). \tag{9}$$

Together with (8) and (9) we obtain

$$b_1 + b_2 + \cdots + b_6 \geq 6 \cdot \frac{2n+1}{3} = 4n + 2,$$

which is in contradiction with (2).

Therefore $n \equiv 0 \pmod 3$, and (8) implies $b_6 > \frac{2n}{3}$. Since b_6 and $\frac{2n}{3}$ are both integers,

$$b_6 \geq \frac{2n}{3} + 1. \tag{10}$$

Together with (4) and (10) we obtain

$$b_1 + b_2 + \cdots + b_6 \geq 5 \cdot \frac{2n}{3} + \frac{2n}{3} + 1 = 4n + 1. \tag{11}$$

Compare with (2), we see that (11) should be an equality, so are (5) and (10). Hence $b_6 = \frac{2n}{3} + 1$.

Consider

$$\sum_{1 \leq i < j \leq 5} x_{ij} \geq \sum_{1 \leq i < j \leq 5} \frac{2n+1}{5} = 2(2n+1) = 4n + 2. \tag{12}$$

Since one contestant solved $P_1, P_2, \ldots, P_5$, whereas b_6 contestants solved exactly three problems of $P_1, P_2, \ldots, P_5$, and $n - 1 - b_6$ contestants solved exactly four problems of $P_1, P_2, \ldots, P_5$, we have

$$\sum_{1 \le i < j \le 5} x_{ij} = C_5^2 + b_6 \cdot C_3^2 + (n - 1 - b_6)C_4^2$$

$$= 6n + 4 - 3b_6 = 4n + 1,$$

which is in contradiction with (12). This completes the proof.

【Score Situation】 This particular problem saw the following distribution of scores among contestants: 56 contestants scored 7 points, 6 contestants scored 6 points, 2 contestants scored 5 points, 15 contestants scored 4 points, 13 contestants scored 3 points, 57 contestants scored 2 points, 39 contestants scored 1 point, and 325 contestants scored 0 point. The average score of this problem is 1.345, indicating that it was relatively challenging.

  Among the top five teams in the team scores, the scores of this problem are as follows: the China team scored 42 points (with a total team score of 235 points), the United States team scored 35 points (with a total team score of 213 points), the Russia team scored 31 points (with a total team score of 212 points), the Iran team scored 22 points (with a total team score of 201 points), and the South Korea team scored 19 points (with a total team score of 200 points).

  The gold medal cutoff for this IMO was set at 35 points (with 42 contestants earning gold medals), the silver medal cutoff was 23 points (with 79 contestants earning silver medals), and the bronze medal cutoff was 12 points (with 128 contestants earning bronze medals).

  In this IMO, a total of 16 contestants achieved a perfect score of 42 points.

Problem 2.9 (IMO 48-3, proposed by Russia). In a mathematical competition some competitors are friends. Friendship is always mutual. Call a group of competitors a *clique* if each two of them are friends. (In particular, any group of fewer than two competitors is a clique.) The number of members of a clique is called its *size*.

  Given that, in this competition, the largest size of a clique is even, prove that the competitors can be arranged in two rooms such that the largest size of a clique contained in one room is the same as the largest size of a clique contained in the other room.

Proof. We provide an algorithm to distribute competitors.

  Denote the rooms A and B. At some initial stage, we move one person at a time from one room to the other one. We reach the target by several adjustments. In every step of the algorithm, let A and B be the sets of competitors in room A and room B, and $C(A)$ and $C(B)$ be the largest size of cliques in room A and room B.

Step 1: Let M be the largest size of clique of all competitors, $|M| = 2m$.

Move all members of M to room A, and the left ones to room B. Since M is the largest clique of all competitors, $C(A) = |M| \geq C(B)$.

Step 2: If and only if $C(A) > C(B)$, move one person from room A to room B. (In view of $C(A) > C(B)$, we see that $C(A) \neq \varnothing$.)

After every operation is done, $C(A)$ decreases by 1 while $C(B)$ increases by 1 at most. These operations will not end until

$$C(A) \leq C(B) \leq C(A) + 1.$$

At that time, $C(A) = |A| \geq m$. (Otherwise there are at least $m + 1$ members of M in room B and at most $m - 1$ members in room A. Then $C(B) - C(A) \geq (m + 1) - (m - 1) = 2$, which is impossible.)

Step 3: Denote $K = C(A)$.

If $C(B) = K$, then we are done; or else $C(B) = K + 1$. From the above discussion, $K = |A| = |A \cap M| \geq m$ and $|B \cap M| \leq m$.

Step 4: If there is a largest size clique C in room B and a competitor $x \in B \cap M$ but $x \notin C$, then move x to room A, and we are done.

In fact, after the operation, there are $K + 1$ members of M in room A, so $C(A) = K + 1$. Since $x \notin C$, whose removal does not destroy C, we have $C(B) = |C|$. Therefore, $C(A) = C(B) = K + 1$.

If such competitor x does not exist, then each largest clique of room B contains $B \cap M$ as a subset. In this case, we do step 5. $\hspace{1em}(*)$

Step 5: Choose any of the largest clique $C(|C| = K + 1)$ in room B and move a member of $C \backslash M$ to room A.

(In view of $|C| = K + 1 > m \geq |B \cap M|$, we know $C \backslash M \neq \varnothing$.)

Since we only move one person at a time from room B to room A, so $C(B)$ decreases by 1 at most. At the end of this step, $C(B) = K$.

Now, there is a clique $A \cap M$ in A with $|A \cap M| = K$. So $C(A) \geq K$. We prove $C(A) = K$ as follows.

Let Q be any clique of room A. We only need to show $|Q| \leq K$. In fact, the members of room A can be classified into two types:

(1) Members of M. Since M is a clique, they and members of $B \cap M$ are friends.
(2) The members move from room B to room A at step 5. By $(*)$, they are friends of $B \cap M$.

So, every member of Q and members of $B \cap M$ are friends. What's more, Q and $B \cap M$ both are cliques, so is $Q \cup (B \cap M)$.

Because M is the largest size clique of all competitors,

$$|M| \geq |Q \cup (B \cap M)| = |Q| + |B \cap M|$$
$$= |Q| + |M| - |A \cap M|,$$

so $|Q| \leq |A \cap M| = K$.

Therefore, after these 5 steps, $C(A) = C(B) = K$.

【Score Situation】 This particular problem saw the following distribution of scores among contestants: 2 contestants scored 7 points, 1 contestant scored 6 points, 1 contestant scored 5 points, 3 contestants scored 4 points, 11 contestants scored 3 points, 23 contestants scored 2 points, 42 contestants scored 1 point, and 437 contestants scored 0 point. The average score of this problem is 0.304, indicating that it was extremely difficult.

Among the top five teams in the team scores, the scores of this problem are as follows: the Russia team scored 12 points (with a total team score of 184 points), the China team scored 17 points (with a total team score of 181 points), the South Korea team scored 9 points (with a total team score of 168 points), the Vietnam team scored 10 points (with a total team score of 168 points), and the United States team scored 6 points (with a total team score of 155 points).

The gold medal cutoff for this IMO was set at 29 points (with 39 contestants earning gold medals), the silver medal cutoff was 21 points (with 83 contestants earning silver medals), and the bronze medal cutoff was 14 points (with 131 contestants earning bronze medals).

In this IMO, no contestant achieved a perfect score of 42 points.

Problem 2.10 (IMO 56-6, proposed by Australia). A sequence $a_1, a_2, \ldots$ of integers satisfies the following conditions:

(i) $1 \leq a_j \leq 2015$ for all $j \geq 1$;
(ii) $k + a_k \neq l + a_l$ for all $1 \leq k < l$.

Prove that there exist two positive integers b and N such that

$$\left| \sum_{j=m+1}^{n} (a_j - b) \right| \leq 1007^2$$

for all integers m and n satisfying $n > m \geq N$.

Proof. Let $s_n = n + a_n$. Then $n + 1 \leq s_n \leq n + 2015$, and all the s_i's are distinct.

Let $S = \{s_1, s_2, \ldots\}$. We first prove that $M = \mathbf{N}^* \backslash S$ is finite and $1 \leq |M| \leq 2015$.

Clearly $1 \in M$. Assume that there are more than 2015 elements in M, say, $m_1 < m_2 < \cdots < m_{2016}$. Pick an integer $n > m_{2016}$. Then

$$\{s_1, s_2, \ldots, s_n\} \subset \{1, 2, \ldots, n + 2015\},$$

$$\{m_1, m_2, \ldots, m_{2016}\} \subset \{1, 2, \ldots, n + 2015\}.$$

The sets $\{s_1, s_2, \ldots, s_n\}$ and $\{m_1, m_2, \ldots, m_{2016}\}$ are disjoint by the definition of M. However, $n + 2016 > n + 2015$, a contradiction. So M is finite and $1 \le |M| \le 2015$.

Let $b = |M|$ and pick an integer $N > \max |M|$. We prove that these numbers satisfy the requirement in the problem.

Clearly b and N are well defined integers, and we proved that $1 \le b \le 2015$. For any $n \ge N$, we have the following partition:

$$\{1, 2, \ldots, n + 2015\} = \{s_1, s_2, \ldots, s_n\} \cup M \cup C_n, \tag{1}$$

where $C_n = \{1, 2, \ldots, n + 2015\} \setminus (\{s_1, s_2, \ldots, s_n\} \cup M)$ and $|C_n| = 2015 - b$.

For $j \ge n + 1$, we see that $s_j = j + a_j \ge n + 1 + 1 = n + 2$. If C_n is not empty, we must have $n + 2 \le \min C_n$; otherwise, $\min C_n$ is not in S, nor in M, which is impossible. So

$$C_n \subseteq \{n + 2, n + 3, \ldots, n + 2015\}. \tag{2}$$

Note that (2) trivially holds when C_n is empty. Compute the summation of the elements on both sides of (1). Writing $\sigma(X)$ the sum of the elements in a set X, we have

$$\sum_{i=1}^{n+2015} i = \sum_{j=1}^{n} s_j + \sigma(M) + \sigma(C_n).$$

Substituting $s_j = j + a_j$, we get

$$\sum_{i=n+1}^{n+2015} i = \sum_{j=1}^{n} a_j + \sigma(M) + \sigma(C_n).$$

For $n > m \ge N$, substituting m for n to get a new equation and taking the difference, we obtain

$$\sum_{i=n+1}^{n+2015} i - \sum_{i=m+1}^{m+2015} i = \sum_{j=m+1}^{n} a_j + \sigma(C_n) - \sigma(C_m).$$

Subtracting $(n - m)b$ from both sides and simplifying, we get

$$\sum_{j=m+1}^{n} (a_j - b) = (2015 - b)(n - m) + \sigma(C_m) - \sigma(C_n). \tag{3}$$

By (2) and $|C_n| = 2015 - b$ we have

$$\sum_{i=n+2}^{n+2016-b} i \leq \sigma(C_n) \leq \sum_{i=n+b+1}^{n+2015} i,$$

i.e., $(n + 1009 - \frac{b}{2})(2015 - b) \leq \sigma(C_n) \leq (n + 1008 + \frac{b}{2})(2015 - b)$.

The same estimate holds with n replaced by m. Using these estimates in (3), we get

$$\sum_{j=m+1}^{n} (a_j - b) \leq (2015 - b)(n - m) + \left(m + 1008 + \frac{b}{2}\right)(2015 - b)$$

$$- \left(n + 1009 - \frac{b}{2}\right)(2015 - b)$$

$$= (2015 - b)(b - 1) \leq 1007^2$$

and

$$\sum_{j=m+1}^{n} (a_j - b) \geq (2015 - b)(n - m) + \left(m + 1009 - \frac{b}{2}\right)(2015 - b)$$

$$- \left(n + 1008 + \frac{b}{2}\right)(2015 - b)$$

$$= -(2015 - b)(b - 1) \geq -1007^2.$$

Thus, $|\sum_{j=m+1}^{n} (a_j - b)| \leq 1007^2$.

【Score Situation】 This particular problem saw the following distribution of scores among contestants: 11 contestants scored 7 points, 7 contestants scored 6 points, 3 contestants scored 5 points, 3 contestants scored 4 points, 6 contestants scored 3 points, 15 contestants scored 2 points, 11 contestants scored 1 point, and 521 contestants scored 0 point. The average score of this problem is 0.355, indicating that it was extremely difficult.

Among the top five teams in the team scores, the scores of this problem are as follows: the United States team scored 13 points (with a total team score of 185 points), the China team scored 27 points (with a total team score of 181 points), the South Korea team scored 19 points (with a total team score of 161 points), the North Korea team scored 8 points (with a total team score of 156 points), and the Vietnam team scored 2 points (with a total team score of 151 points).

The gold medal cutoff for this IMO was set at 26 points (with 39 contestants earning gold medals), the silver medal cutoff was 19 points (with 100 contestants earning silver medals), and the bronze medal cutoff was 14 points (with 143 contestants earning bronze medals).

In this IMO, only one contestant achieved a perfect score of 42 points, namely Zhuo Qun Alex Song from Canada.

Problem 2.11 (IMO 58-5, proposed by Russia). An integer $N \geq 2$ is given. A collection of $N(N + 1)$ soccer players, no two of whom are of the same height, stand in a row. Sir Alex wants to remove $N(N - 1)$ players from this row, leaving a new row of $2N$ players in which the following N conditions hold:

(1) no one stands between the two tallest players;
(2) no one stands between the third and fourth tallest players;

$$\vdots$$

(N) no one stands between the two shortest players.
Show that this is always possible.

Proof 1. Split the row of $N(N + 1)$ players into N blocks with $N + 1$ consecutive people each. Consider the second highest player among each block. Let B_1 be the highest player among these N players and let A_1 be the highest player in the block of B_1. Remove the remaining $N - 1$ players in the block of A_1, B_1, and remove the $N - 1$ highest players in each other block as well.

Note that the remaining $N(N - 1)$ players other than A_1 and B_1 are distributed in $N - 1$ blocks with N players in each block, and they are all shorter than B_1. Repeat the same procedure for these $N(N - 1)$ players. We can choose A_2 and B_2 in some block and remove $2(N - 2)$ players afterward. There are $(N - 1)(N - 2)$ players left.

Repeat the same procedure, we can eventually choose $2N$ players: $A_1, B_1, A_2, B_2, \ldots, A_N, B_N$. Their heights are in decreasing order. Since A_i and B_i are in the same block initially, no one stands between them. This completes our proof.

Proof 2. Split the $N(N + 1)$ players into N groups by height: the first group has the $N + 1$ tallest ones, the second group has the next $N + 1$ tallest, and so on, up to the N-th group with the $N + 1$ shortest people. Now scan the original row from left to right, stopping as soon as you have scanned two people from the same group, say player A_1 and A_2, are in group t_1 with A_1 on the left of A_2.

Let A_1 and A_2 stay in the row, the other people on the left of A_2 leave the row, and the other people of group t_1 leave the row as well. Note that the remaining people on the right of A_2 belong to the other $N - 1$ groups.

Since each group has at most one person leaving the row, each of the $N-1$ groups has at least N people in the row.

We now scan from A_2 to the right until we find two other players in the same group. Denote these two players by A_3 and A_4 in group t_2. Let A_3 and A_4 stay in the row, the other people between A_2 and A_4 leave the row, and the other people of group t_2 leave the row as well.

Keep doing this procedure. Suppose we have chosen $A_1, A_2, \ldots, A_{2k-1}$, and A_{2k} which are ordered from left to right in the row, A_1 and A_2 in group t_1, A_3 and A_4 in group $t_2, \ldots, A_{2k-1}$ and A_{2k} in group t_k, and the remaining people on the right of A_{2k} all belong to the remaining $N-k$ groups with at least $N-k+1$ people in each of these groups.

If $k \leq N-1$, we scan from A_{2k} to the right until another two players in the same group are found. Denote these two players by A_{2k+1} and A_{2k+2} with A_{2k+1} on the left of A_{2k+2}, say they are in group t_{k+1}. Let them stay in the row, the other people between A_{2k} and A_{2k+2} leave the row, and the other people of group t_{k+1} leave the row as well.

Finally, we have chosen $2N$ players $A_1, A_2, \ldots, A_{2N}$, ordered from left to right in the row, with two players from each group. Since the two players from the same group are consecutive in this sequence, no other players stand between them. This completes the proof.

【Score Situation】 This particular problem saw the following distribution of scores among contestants: 59 contestants scored 7 points, 1 contestant scored 6 points, 2 contestants scored 5 points, no contestant scored 4 points, 9 contestants scored 3 points, 47 contestants scored 2 points, 46 contestants scored 1 point, and 451 contestants scored 0 point. The average score of this problem is 0.969, indicating that it was extremely difficult.

Among the top five teams in the team scores, the scores of this problem are as follows: the South Korea team scored 22 points (with a total team score of 170 points), the China team scored 19 points (with a total team score of 159 points), the Vietnam team scored 21 points (with a total team score of 155 points), the United States team scored 23 points (with a total team score of 148 points), and the Iran team scored 17 points (with a total team score of 142 points).

The gold medal cutoff for this IMO was set at 25 points (with 48 contestants earning gold medals), the silver medal cutoff was 19 points (with 90 contestants earning silver medals), and the bronze medal cutoff was 16 points (with 153 contestants earning bronze medals).

In this IMO, no contestant achieved a perfect score of 42 points.

Problem 2.12 (IMO 59-3, proposed by Iran). An *anti-Pascal triangle* is an equilateral triangular array of numbers such that, except for the numbers in the bottom row, each number is the absolute value of the difference of the two numbers immediately below it. For example, the following array is an anti-Pascal triangle with four rows which contains every integer from 1 to 10.

$$4$$

$$2 \qquad 6$$

$$5 \qquad 7 \qquad 1$$

$$8 \qquad 3 \qquad 10 \qquad 9$$

Does there exist an anti-Pascal triangle with 2018 rows which contains every integer from 1 to $1 + 2 + \cdots + 2018$?

Solution 1. The answer is negative.

Assume on the contrary that such an anti-Pascal triangle exists. Set $N = 1 + 2 + \cdots + 2018$ and denote the jth number from left to right of the ith row by $a_{i,j}$. Let $a_1 = a_{1,1}$, denote the two numbers immediately below a_1 by a_2 and b_2 with $a_2 > b_2$, and denote the two numbers immediately below a_2 by a_3 and b_3 with $a_3 > b_3$, and so on, we denote the two numbers immediately below a_i by a_{i+1} and b_{i+1} with $a_{i+1} > b_{i+1}$, $i = 1, 2, \ldots, 2017$.

By the definition of an anti-Pascal triangle, $a_i = a_{i+1} - b_{i+1}$ for $i = 1, 2, \ldots, 2017$. Thus,

$$a_{2018} = a_1 + b_2 + b_3 + \cdots + b_{2018}.$$

Clearly, $a_1, b_2, b_3, \ldots, b_{2018}$ are 2018 distinct positive integers, and we have

$$a_1 + b_2 + b_3 + \cdots + b_{2018} \geq 1 + 2 + 3 + \cdots + 2018 = N.$$

Since $a_{2018} \leq N$, we obtain $a_{2018} = N$, and

$$\{a_1, b_2, b_3, \ldots, b_{2018}\} = \{1, 2, 3, \ldots, 2018\}.$$

Let $a_{2018} = a_{2018,j}$. By symmetry, we may assume $j \leq 1009$ without loss of generality. Since

$$a_{2018}, b_{2018} \in \{a_{2018,1}, a_{2018,2}, \ldots, a_{2018,1010}\},$$

the numbers $a_1, a_2, \ldots, a_{2018}$ and $b_2, b_3, \ldots, b_{2018}$ all belong to the following set S:

$$S = \{a_{i,j} | j \leq 1010\}.$$

Consider the remaining numbers

$$T = \{a_{i,j} \mid 1011 \leq i \leq 2018, 1011 \leq j \leq i\}.$$

Let $c_{1011} = a_{1011,1011}$. For $i = 1011, 1012, \ldots, 2017$, denote the two numbers immediately below c_i by c_{i+1} and d_{i+1} with $c_{i+1} > d_{i+1}$. Since $c_i = c_{i+1} - d_{i+1}$ for $i = 1011, 1012, \ldots, 2017$, we have

$$c_{2018} = c_{1011} + d_{1012} + d_{1013} + \cdots + d_{2018}.$$

Since $c_{1011}, d_{1012}, d_{1013}, \ldots, d_{2018}$ are in T, not in the set S, they are all larger than 2018. Thus $c_{1011} + d_{1012} + d_{1013} + \cdots + d_{2018} \geq 2019 + 2020 + \cdots + 3026 = 2542680 > N = 2037171$. This contradicts the fact $c_{2018} \leq N$, and shows that the required anti-Pascal triangle does not exist.

Solution 2. The answer is negative.

Assume on the contrary that such an anti-Pascal triangle exists. Set $N = \frac{1}{2} \times 2018 \times 2019$. We call $1, 2, \ldots, 2018$ small numbers and N, $N-1, \ldots, N-2018$ big numbers. Note that the two numbers immediately below a big number are formed by one big number and one small number.

(1) We claim that there is exactly one small number on each row.

Let a_i and b_i be the maximum and minimum number on the ith row, respectively. For $1 \leq i \leq 2017$, assume that the two numbers immediately below a_i are b and c with $b > c$. Then $a_i = b - c \leq a_{i+1} - b_{i+1}$.

Summing the above inequalities over $i = 1, 2, \ldots, 2017$, we see that

$$a_1 \leq a_{2018} - \sum_{i=2}^{2018} b_i.$$

Clearly, $a_1, b_2, b_3, \ldots, b_{2018}$ are distinct positive integers. We have

$$a_{2018} \geq a_1 + \sum_{i=2}^{2018} b_i \geq \sum_{i=1}^{2018} i = N.$$

Since $a_{2018} \leq N$, we obtain $a_{2018} = N$, and

$$\{a_1, b_2, b_3, \ldots, b_{2018}\} = \{1, 2, \ldots, 2018\}.$$

(2) All big number are in the last 80 rows.

For $i \leq 1938$,

$$a_i \leq a_{2018} - \sum_{j=i+1}^{2018} b_j \leq N - \sum_{j=1}^{2018-i} j \leq N - \sum_{j=1}^{80} j = N - 3240,$$

so a_i is not a big number. Thus, there is not a big number in the first 1938 rows.

(3) We'll estimate the number of big numbers to get a contradiction.

If there are two consecutive big numbers in the last row, then the number above them is a small number. Since there is only one small number in the second last row, there is at most one pair of consecutive big numbers. Thus, the number of big numbers in the last row does not exceed $2 + \lceil \frac{2016}{2} \rceil = 1010$.

The big numbers which are not in the last row have a small number immediately below it. Since big numbers are in the last 80 rows and there is only one small number in each row, we conclude that the number of big numbers not in the last row is at most $79 \times 2 = 158$. Thus, the number of big numbers is no more than 1168, which contradicts the fact that there are exactly 2019 big numbers.

Thus, the required anti-Pascal triangle does not exist.

【Score Situation】 This particular problem saw the following distribution of scores among contestants: 11 contestants scored 7 points, no contestant scored 6 points, 1 contestant scored 5 points, 4 contestants scored 4 points, 14 contestants scored 3 points, 9 contestants scored 2 points, 7 contestants scored 1 point, and 548 contestants scored 0 point. The average score of this problem is 0.278, indicating that it was extremely difficult.

Among the top five teams in the team scores, the scores of this problem are as follows: the United States team scored 14 points (with a total team score of 212 points), the Russia team scored 11 points (with a total team score of 201 points), the China team scored 17 points (with a total team score of 199 points), the Ukraine team scored 12 points (with a total team score of 186 points), and the Thailand team scored 3 points (with a total team score of 183 points).

The gold medal cutoff for this IMO was set at 31 points (with 48 contestants earning gold medals), the silver medal cutoff was 25 points (with 98 contestants earning silver medals), and the bronze medal cutoff was 16 points (with 143 contestants earning bronze medals).

In this IMO, only two contestants achieved a perfect score of 42 points, namely Agnijo Banerjee from the United Kingdom and James Lin from the United States.

Problem 2.13 (IMO 64-5, proposed by the Netherlands). Let n be a positive integer. A *Japanese triangle* consists of $1 + 2 + \cdots + n$ circles arranged in an equilateral triangular shape such that for each $i = 1, 2, \ldots, n$, the ith row contains exactly i circles, exactly one of which is colored red. A *ninja path* in a Japanese triangle is a sequence of n circles obtained by starting in the top row, then repeatedly going from a circle to one of the two circles immediately below it and finishing in the bottom row. Figure 2.4 is an example of a Japanese triangle with $n = 6$, along with a ninja path in that triangle containing two red circles.

In terms of n, find the greatest k such that in each Japanese triangle there is a ninja path containing at least k red circles.

Solution. The largest integer k is $\lfloor \log_2 n \rfloor + 1$. Here $\lfloor x \rfloor$ represents the greatest integer less than or equal to x. Let $N = \lfloor \log_2 n \rfloor$. Then $2^N \le n \le 2^{N+1} - 1$.

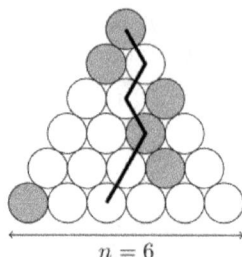

Figure 2.4 A Japanese Triangle with $n = 6$

First, we provide a construction that ensures each ninja path passes through at most $N + 1$ red circles. For row $i = 2^a + b$, where $0 \le a \le N$ and $0 \le b \le 2^a - 1$, color the $(2b + 1)$th circle red, as shown in Figure 2.5.

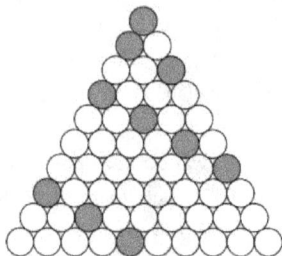

Figure 2.5 A Japanese Triangle with $n = 10$

Thus, for any $a = 0, 1, \ldots, N$, each ninja path in the rows 2^a, $2^a + 1, \ldots, 2^{a+1} - 1$ passes through at most one red circle. Therefore, each ninja path passes through at most $N + 1$ red circles.

Now, we prove that in every Japanese triangle, there exists a ninja path passing through at least $N + 1$ red circles. In each circle C, label the maximum number of red circles encountered in a ninja path starting from the topmost circle and ending at C, as shown in Figure 2.6.

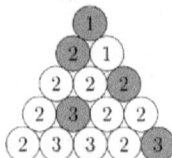

Figure 2.6 A Japanese Triangle with $n = 5$

If C is not red, then the number labeled in C is the maximum of the numbers labeled in the circles immediately upper-left or upper-right of C. If C is red, then the number labeled in C is the maximum of the upper-left or upper-right number plus 1.

Let σ_i be the sum of the numbers labeled in circles in row i. Suppose the numbers in row i are $v_1, v_2, \ldots, v_i$, and v_m is the largest among them. If there is no red circle in the row $i+1$, then the sum of the numbers labeled in the row $i+1$ is at least $(v_1 + v_2 + \cdots + v_i) + v_m$. Therefore, σ_{i+1} is at least $(v_1 + v_2 + \cdots + v_i) + v_m + 1$. We prove the following lemma.

Lemma. *For any $0 \leq j \leq N$, there holds that $\sigma_{2^j} \geq j \cdot 2^j + 1$.*

Proof of the Lemma. We use induction on j. For $j = 0$, the conclusion is evident. Assume $\sigma_{2^j} \geq j \cdot 2^j + 1$. Then the maximum of the numbers labeled in the row 2^j is at least $j + 1$. For any $k \geq 2^j$, there is a labeled number in the row k at least $j + 1$. Thus

$$\sigma_{k+1} \geq \sigma_k + (j+1) + 1 = \sigma_k + (j+2),$$

$$\sigma_{2^{j+1}} \geq \sigma_{2^j} + 2^j(j+2) \geq j \cdot 2^j + 1 + 2^j(j+2) = (j+1)2^{j+1} + 1.$$

Therefore, for $j = N$, by the lemma, there exists a circle in row 2^N with a labeled number at least $N + 1$. Hence, there exists a ninja path passing through at least $N + 1$ red circles.

So $\lfloor \log_2 n \rfloor + 1$ is the maximum of k.

【Score Situation】 This particular problem saw the following distribution of scores among contestants: 118 contestants scored 7 points, 9 contestants scored 6 points, 13 contestants scored 5 points, 4 contestants scored 4 points, 52 contestants scored 3 points, 174 contestants scored 2 points, 29 contestants scored 1 point, and 219 contestants scored 0 point. The average score of this problem is 2.417, indicating that it had a certain level of difficulty.

Among the top five teams in the team scores, the scores of this problem are as follows: the China team scored 42 points (with a total team score of 240 points), the United States team scored 37 points (with a total team score of 222 points), the South Korea team scored 33 points (with a total team score of 215 points), the Romania team scored 24 points (with a total team score of 208 points), and the Canada team scored 38 points (with a total team score of 183 points).

The gold medal cutoff for this IMO was set at 32 points (with 54 contestants earning gold medals), the silver medal cutoff was 25 points (with 90 contestants earning silver medals), and the bronze medal cutoff was 18 points (with 170 contestants earning bronze medals).

In this IMO, a total of five contestants achieved a perfect score of 42 points.

2.3 Summary

In the first 64 IMOs, there were a total of 13 existence problems, as depicted in Figure 2.7. The score details for these problems are presented in Table 2.2. Due to the smaller number of participating teams and missing contestant score information in early IMOs, there are several blanks in Table 2.2.

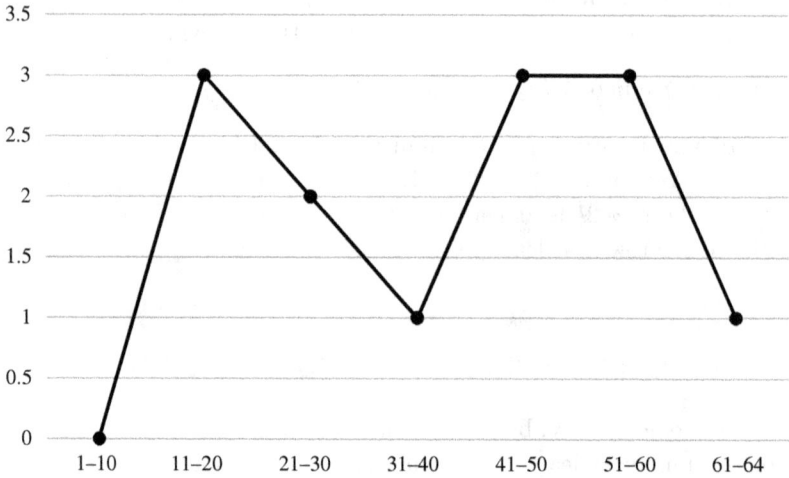

Figure 2.7 Numbers of Existence Problems in the First 64 IMOs

These 13 problems were proposed by 10 countries. Russia, Iran, and the Netherlands each contributed two problems.

From Table 2.2, it can be observed that in the first 64 IMOs, there were six existence problems with an average score of 0–1 point; one problem with an average score of 1–2 points; three problems with an average score of 2–3 points; two problems with an average score of 3–4 points; one problem with an average score above 4 points. Among these problems, the one with the lowest average score is Problem 2.12 (IMO 59-3), proposed by Iran. Overall, the existence problems were relatively challenging, and the contestants scored relatively low.

In the 24th–64th IMOs, there were a total of 10 existence problems. Among these, five had an average score of 0–1 point; one had an average score of 1–2 points; three had an average score of 2–3 points; one had an average score of 3–4 points. Further analysis of the problem numbers of these 10 existence problems, as shown in Table 2.3, reveals that these problems frequently appeared as the 3rd/6th problem.

Table 2.2 Score Details of Existence Problems in the First 64 IMOs

Problem	2.1	2.2	2.3	2.4	2.5	2.6	2.7	2.8
Full points	6	5	8	7	7	7	7	7
Average score	4.943	3.061	0.917	2.150	2.411	3.743	0.877	1.345
Top five mean		4.050	1.975		4.733	6.467	3.233	4.967
6th–15th mean			0.908		2.800	5.767	1.233	3.467
16th–25th mean					2.017	5.183	1.700	2.050
Problem number in the IMO	12-4	14-1	20-6	24-5	26-4	38-4	42-3	46-6
Proposing country	Czechoslovakia	The Soviet Union	The Netherlands	Poland	Mongolia	Iran	Germany	Romania

Problem	2.9	2.10	2.11	2.12	2.13
Full points	7	7	7	7	7
Average score	0.304	0.355	0.969	0.278	2.417
Top five mean	1.800	2.300	3.400	1.900	5.800
6th–15th mean	0.500	1.083	2.333	1.017	4.550
16th–25th mean	0.583	0.450	1.815	0.483	4.100
Problem number in the IMO	48-3	56-6	58-5	59-3	64-5
Proposing country	Russia	Australia	Russia	Iran	The Netherlands

Note. Top five mean = Total score of the top five teams ÷ Total number of contestants from the top five teams,
6th–15th mean = Total score of the 6th–15th teams ÷ Total number of contestants from the 6th–15th teams,
16th–25th mean = Total score of the 16th–25th teams ÷ Total number of contestants from the 16th–25th teams.

Table 2.3   Numbers of Existence Problems in the 24th–64th IMOs

Existence Problems	Problem Number			Number of Problems in the First 64 IMOs
	1, 4	2, 5	3, 6	
Existence problems	2	3	5	13

From Table 2.2, it is observable that in the existence problems, the average score of the top five teams generally exceeds the average score of the problem by 2.5 points, the average score of the 6th–15th teams typically surpasses the average score by 2 points, and the average score of the 16th–25th teams usually exceeds the average score by 1 point.

However, it can also be observed that there is an existence problem where the average score is higher than the average score of the 16th–25th teams, such as Problem 2.5 (IMO 26-4). This phenomenon is due to the smaller number of participating teams in early IMOs. It was not until the 30th IMO in 1989 that the number of participating teams exceeded 50. Therefore, it is common to see situations where the average score is close to or even higher than the average score of the 16th–25th teams during this period.

Chapter 3

Extremal Combinatorial Problems

Extremum problems are a focal point in mathematical competitions. In algebraic extremum problems, variables are generally continuous, and one can employ inequalities or functional methods to solve these problems. However, in combinatorial extremum problems, the variables are typically discrete, such as integers, sets, and graphs. As a result, structures of combinatorial extremum problems are complex, making it challenging to directly express corresponding functional relationships in analytical form, and thus difficult to solve using algebraic methods.

Typically, solving combinatorial extremum problems can be divided into three steps:

(1) Employ various methods to hypothesize the maximum or minimum value.
(2) Construct an example that fits the problem's conditions, demonstrating that this value can be obtained.
(3) Prove that this value is the extremum, meaning that no larger or smaller value can be attained.

In the first 64 IMOs, there had been a total of four extremal combinatorial problems, accounting for approximately 4.6% of all combinatorics problems. The statistical distribution of these problems in the previous IMOs is presented in Table 3.1.

Table 3.1 Numbers of Extremal Combinatorial Problems in the First 64 IMOs

Content	Session							Total
	1–10	11–20	21–30	31–40	41–50	51–60	61–64	
Extremal combinatorial problems	0	2	0	2	0	0	0	4
Combinatorics problems	7	11	16	16	11	19	7	87
Percentage of extremal combinatorial problems among combinatorics problems	0.0%	18.2%	0.0%	12.5%	0.0%	0.0%	0.0%	4.6%

Because extremal combinatorial problems are often classified to other areas (such as combinatorial geometry and operation problems), they appear less frequently in this statistical analysis.

It is important to note that for each problem, the solutions are followed by information on the scores, including the number of contestants in each score range, the average score, and the scores of the top five teams. However, early IMOs often lacked information on contestant scores, so the number of contestants in each score range only represents the counted number of contestants, and some problems lack scores of the top five teams.

3.1 Common Theorems, Formulas, and Methods

3.1.1 *Common methods*

(1) *Estimation method*

The estimation method is the most common way to solve extremal combinatorial problems, which requires first experimenting with simple scenarios to guess the optimal value before proving it.

Example 3.1. Let S be a set consisting of positive integers not greater than 15, such that any two disjoint nonempty subsets of S have different sums of their elements. Determine the maximum possible value of the sum of elements in S.

Solution. We first claim that $|S| \leq 5$. Suppose on the contrary that $|S| = n \geq 6$, the number of nonempty subsets of S with no more than four elements is

$$C_n^1 + C_n^2 + C_n^3 + C_n^4 \geq C_6^1 + C_6^2 + C_6^3 + C_6^4 = 56.$$

Since $12 + 13 + 14 + 15 = 54 < 56$, there exist two subsets with the same sum of elements by the pigeonhole principle. Deleting their common elements, we obtain a contradiction. Thus $|S| \leq 5$.

Next, we construct a five-element set S satisfying the required property with a maximum sum of elements. We use the greedy algorithm to pick elements of S as large as possible. Choose 15, 14, and 13. Then $12 \notin S$ since $15 + 12 = 13 + 14$.

Now choose 11. Then $10 \notin S$ and $9 \notin S$ since $15 + 10 = 14 + 11$ and $15 + 9 = 13 + 11$. Now choose 8. It is not hard to verify that $S = \{15, 14, 13, 11, 8\}$ satisfies the required property, with the sum of elements $15 + 14 + 13 + 11 + 8 = 61$.

Finally, we claim 61 is the maximum possible value. Since $15 + 14 + 13 + 12 < 61$, the sum of elements of S is less than 61 if $|S| \leq 4$. Consider $|S| = 5$, say $S = \{a, b, c, d, e\}$ with $a > b > c > d > e$. Clearly $a + b + c + d \leq 15 + 14 + 13 + 11$. If $e \leq 8$, then we are done.

Let $e \geq 9$ and suppose $a + b + c + d + e > 61$. By the above greedy algorithm, $(a, b, c, d) \neq (15, 14, 13, 11)$ and the maximum is obtained by $(15, 14, 13, 10, 9)$, which is 61 (in fact it does not satisfy the condition).

We conclude that the maximum possible value of the sum of elements in S is 61.

(2) *Method of adjustment*

When the optimum value is not reached, adjusting the known variable can result in larger or smaller values, to conclude that the original value is not the optimum value.

Example 3.2. Write 201 as the sum of four different positive integers x_1, x_2, x_3, and x_4. Let

$$s = x_1 x_2 + x_1 x_3 + x_1 x_4 + x_2 x_3 + x_2 x_4 + x_3 x_4.$$

Find the maximum possible value of s.

Solution. Since there are only finitely many ways to represent 201 as the sum of 4 different positive integers, the maximum can be achieved in some way.

Let $x_1 < x_2 < x_3 < x_4$ be positive integers satisfying $x_1 + x_2 + x_3 + x_4 = 201$, which achieves the maximum of s. We claim that x_1, x_2, x_3, and x_4 have the following properties:

(i) $x_{i+1} - x_i \leq 2 (i = 1, 2, 3)$.

For otherwise, there exists some i with $x_{i+1} - x_i \geq 3$. For example, let $i = 1$. We may write s as $s = x_1 x_2 + (x_1 + x_2)(x_3 + x_4) + x_3 x_4$. Set $x_1' = x_1 + 1$, $x_2' = x_2 - 1$, and $x_i' = x_i (i = 3, 4)$ with $s' = x_1' x_2' + (x_1' + x_2')(x_3' + x_4') + x_3' x_4'$. We have $x_1' < x_2' < x_3' < x_4'$ and $x_1' + x_2' + x_3' + x_4' = 201$. Since

$$s' - s = x_1' x_2' - x_1 x_2 + [(x_1' + x_2') - (x_1 + x_2)](x_3 + x_4)$$

$$= (x_1 + 1) \cdot (x_2 - 1) - x_1 x_2 = x_2 - x_1 - 1 > 0,$$

we obtain $s' > s$, which contradicts the maximality of s.

(ii) $x_{i+1} - x_i = 2$ holds for at most one i.

For otherwise, there are i and j with $1 \leq i < j \leq 3$, such that $x_{i+1} - x_i = 2$ and $x_{j+1} - x_j = 2$. Set $x_i' = x_i + 1$ and $x_{j+1}' = x_{j+1} - 1$. Replace x_i and x_j with x_i' and x_j', with the other numbers unchanged. It is proved similarly as above that s increases, which is a contradiction.

(iii) $x_{i+1} - x_i = 2$ holds for some i.

If any two adjacent numbers of x_1, x_2, x_3, and x_4 differ by 1, then we may assume that the four numbers are k, $k + 1$, $k + 2$, and $k + 3$. It follows that $k + (k + 1) + (k + 2) + (k + 3) = 201$ and $4k = 195$. However, 195 is not divisible by 4, which is absurd. In view of (ii), exactly one pair of two adjacent numbers of x_1, x_2, x_3, and x_4 differ by 2. By considering modulo 4, the only possibility is $k - 1$, $k + 1$, $k + 2$, and $k + 3$, with

$$(k - 1) + (k + 1) + (k + 2) + (k + 3) = 201.$$

Then $4k = 196$, so $k = 49$. Thus $x_1 = 48$, $x_2 = 50$, $x_3 = 51$, and $x_4 = 52$. The maximum value of s is

$$s_{\max} = 48 \times 50 + (48 + 50)(51 + 52) + 51 \times 52 = 15146.$$

(3) *Proof by cases*

Example 3.3. In the calculation of addition of two three-digit numbers
$$\begin{array}{r} abc \\ + \ def \\ \hline ghi \end{array},$$ one finds that a, b, c, d, e, f, g, h, i are exactly the numbers 1, 2, 3, 4, 5, 6, 7, 8, 9 in some order. Find the maximum and the minimum value of ghi.

Solution. Consider the numbers modulo 9, we have $g + h + i = a + b + c + d + e + f - 9k$. It is clear that k is the number of carryovers, so $k < 3$. Since

$$2(g + h + i) = a + b + c + d + e + f + g + h + i - 9k = 45 - 9k,$$

we see that k must be odd. Thus $k = 1$ and $g + h + i = 18$.

The maximum possible value of ghi is 981, and this is possible since
$$\begin{array}{r} 657 \\ + \ 324 \\ \hline 981 \end{array}.$$

Since $g \geq a + d$, we have $g \geq 3$.

If $g = 3$, then $\{a, d\} = \{1, 2\}$. All possible values of ghi are 369, 378, 387, and 396. We also need $c + f \geq 10$. But it is impossible in any of these cases.

If $g = 4$, then the minimum possible value of ghi is 459, and this is possible since
$$\begin{array}{r} 286 \\ + \ 173 \\ \hline 459 \end{array}.$$

(4) *Method of mathematical induction*

Example 3.4. Determine the maximum positive integer m such that an $m \times m$ square can be dissected into 7 rectangles, whose 14 edge lengths consist exactly of 1, 2, 3, 4, 5, 6, 7, 8, 9, 10, 11, 12, 13, 14.

Solution. First we consider a somewhat related problem: assuming that $(a_1, a_2, \ldots, a_{2n})$ is a permutation of $(1, 2, 3, \ldots, 2n)$, we find the maximum of $S_n = a_1 a_2 + a_3 a_4 + \cdots + a_{2n-1} a_{2n}$.

It is easy to see that

$$S_1 = 1 \cdot 2 \quad \text{and}$$

$$S_2 \leq \max \{1 \cdot 2 + 3 \cdot 4, 1 \cdot 3 + 2 \cdot 4, 1 \cdot 4 + 2 \cdot 3\} = 1 \cdot 2 + 3 \cdot 4.$$

We may conjecture that

$$S_n \leq 1 \cdot 2 + 3 \cdot 4 + 5 \cdot 6 + \cdots + (2n - 1) \cdot (2n). \tag{*}$$

For $n = 1$ and 2, inequality (*) holds. Assume that $S_k \leq 1 \cdot 2 + 3 \cdot 4 + \cdots + (2k - 1) \cdot 2k$. Consider $n = k + 1$ and

$$S_{k+1} = a_1 \cdot a_2 + \cdots + a_{2k-1} \cdot a_{2k} + a_{2k+1} \cdot a_{2k+2}. \tag{**}$$

Case 1: $2k + 1$ and $2k + 2$ appear in the same product in (**), say

$$a_{2k+1} \cdot a_{2k+2} = (2k + 1) \cdot (2k + 2).$$

It follows from the inductive hypothesis that

$$S_{k+1} = S_k + (2k+1)(2k+2) \leq 1 \cdot 2 + 3 \cdot 4 + \cdots$$
$$+ (2k-1)(2k) + (2k+1)(2k+2).$$

Case 2: $2k+1$ and $2k+2$ appear in different products in (**). Without loss of generality, we may assume $a_{2k-1} = 2k+1$ and $a_{2k+1} = 2k+2$. Since

$$[a_{2k} \cdot a_{2k+2} + (2k+1)(2k+2)] - [(2k+1)a_{2k} + (2k+2)a_{2k+2}]$$
$$= [(2k+1) - a_{2k+2}][(2k+2) - a_{2k}] > 0,$$

we have

$$S_{k+1} < a_1 \cdot a_2 + \cdots + a_{2k-3} \cdot a_{2k-2} + a_{2k} \cdot a_{2k+2} + (2k+1)(2k+2)$$
$$\leq 1 \cdot 2 + 3 \cdot 4 + \cdots + (2k-1)(2k) + (2k+1)(2k+2).$$

Inequality (*) also holds.

It follows from the above discussion that if an $m \times m$ square can be dissected into n rectangles with side lengths $1, 2, 3, \ldots, 2n-1, 2n$, then the area of the square must satisfy

$$m^2 \leq 1 \cdot 2 + 3 \cdot 4 + \cdots + (2n-1)(2n).$$

Hence

$$m \leq \sqrt{1 \cdot 2 + 3 \cdot 4 + \cdots + (2n-1)(2n)}.$$

In particular, when $n = 7$, we have

$$m \leq \sqrt{1 \cdot 2 + 3 \cdot 4 + \cdots + 13 \cdot 14} = \sqrt{504} < 23.$$

Thus $m \leq 22$.

Note that $22 = 14 + 8 = 13 + 9 = 12 + 10 = 11 + 6 + 5$ and Figure 3.1 give two constructions for $m = 22$.

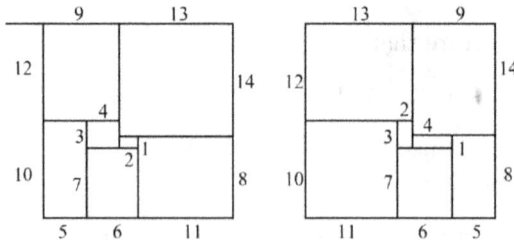

Figure 3.1   Two Constructions for $m = 22$

In conclusion, the maximum m asked is 22.

(5) *Method of counting*

Transform the combinatorial extremum problem into an algebraic problem using counting methods and inequalities before proving it.

Example 3.5. Determine the minimum positive integer n with the following property: for any n numbers in $1, 2, 3, \ldots, 50$, there exist three which are mutually coprime.

Solution. Let $S = \{1, 2, 3, \ldots, 50\}$. Denote by A_i the set of elements of S which are divisible by $i(i = 2, 3)$. We use $|A_2|$ and $|A_3|$ to denote the number of elements of A_2 and A_3, respectively. By the inclusion–exclusion principle,

$$|A_2 \cup A_3| = |A_2| + |A_3| - |A_2 \cap A_3| = \left[\frac{50}{2}\right] + \left[\frac{50}{3}\right] - \left[\frac{50}{2 \times 3}\right]$$

$$= 25 + 16 - 8 = 33.$$

For any three numbers in $A_2 \cup A_3$, two of them belong to A_2 or A_3, and are not coprime. Thus, if we take all the elements of $A_2 \cup A_3$, then there are no three numbers mutually coprime. Hence $n \geq 34$.

On the other hand, let

$$B_1 = \{1, 2, 3, 5, 7, 11, 13, 17, 19, 23, 29, 31, 37, 41, 43, 47\},$$

$$B_2 = \{2^2, 3^2, 5^2, 7^2\}, \quad \text{and} \quad B_3 = \{2 \times 23, 3 \times 17, 5 \times 9\}.$$

There are $16 + 4 + 3 = 23$ numbers in B_1, B_2, and B_3. For any 34 numbers in S, at least $34 - (50 - 23) = 7$ numbers belong to $B_1 \cup B_2 \cup B_3$. By the pigeonhole principle, there are $\left[\frac{7-1}{3}\right] + 1 = 3$ of the 7 numbers that belong to the same set B_1, B_2, or B_3, which are mutually coprime.

In conclusion, the minimum number n asked is 34.

3.2 Problems and Solutions

Problem 3.1 (IMO 18-4, proposed by the United States). Determine, with a proof, the largest number which is the product of positive integers whose sum is 1976.

Solution. The number of different partitions of 1976 into positive integers is finite. Therefore, there is a finite number of associated products among which there must be a largest number.

Suppose we have a partition

$$a_1 + a_2 + \cdots + a_n = 1976$$

such that the associated product $p = a_1 a_2 \ldots a_n$ is maximal. We show that a_i $(i = 1, 2, \ldots, n)$ must satisfy the following properties:

(1) $a_i \leq 4(i = 1, 2, \ldots, n)$.

If some $a_i > 4$, then we replace it with $(a_i - 2) + 2$, and the other numbers remain unchanged. Since

$$2(a_i - 2) = a_i + (a_i - 4) > a_i,$$

this contradicts the maximality of p.

(2) $a_i \neq 1(i = 1, 2, \ldots, n)$.

If $a_1 = 1$, then $(1 + a_2) + a_3 + \cdots + a_n = 1976$ and $(1 + a_2) a_3 \cdots a_n > a_1 a_2 \cdots a_n = p$. Again, this contradicts the maximality of p.

(3) If some $a_i(i = 1, 2, \ldots, n)$ equals 4, then we may replace it with $2 + 2$, which does not change the product p.

It follows from (1), (2), and (3) that p has the form $p = 2^r \cdot 3^s$ with $2r + 3s = 1976$. If $r \geq 3$, then we may replace $2 + 2 + 2$ with $3 + 3$, and the product will increase since $2^3 < 3^2$. So we have $r < 3$.

Now there is a unique solution $1976 = 3 \times 658 + 2$, and the maximum product is $2 \cdot 3^{658}$.

【Score Situation】 This particular problem saw the following distribution of scores among contestants: 50 contestants scored 6 points, 14 contestants scored 5 points, 19 contestants scored 4 points, 10 contestants scored 3 points, 17 contestants scored 2 points, 12 contestants scored 1 point, and 17 contestants scored 0 point. The average score of this problem is 3.755, indicating that it was relatively straightforward.

Among the top five teams in the team scores, the scores of this problem are as follows: the Soviet Union team scored 46 points (with a total team score of 250 points), the United Kingdom team scored 43 points (with a total team score of 214 points), the United States team scored 44 points (with a total team score of 188 points), the Bulgaria team scored 36 points (with a total team score of 174 points), and the Austria team scored 28 points (with a total team score of 167 points).

The gold medal cutoff for this IMO was set at 34 points (with 9 contestants earning gold medals), the silver medal cutoff was 23 points (with 28 contestants earning silver medals), and the bronze medal cutoff was 15 points (with 45 contestants earning bronze medals).

In this IMO, only one contestant achieved a perfect score of 40 points, namely Laurent Pierre from France.

Problem 3.2 (IMO 19-2, proposed by Vietnam). In a finite sequence of real numbers the sum of any seven successive terms is negative, and the sum of any 11 successive terms is positive. Determine the maximum number of terms in the sequence.

Solution 1. We first show that it is not possible for $n \geq 17$. Assume on the contrary that there exists such a sequence $a_1, a_2, \ldots, a_{17}, \ldots, a_n$ for certain $n \geq 17$.

By assumption, we have

$$a_k + a_{k+1} + a_{k+2} + \cdots + a_{k+6} < 0 \; (k \geq 1), a_k + a_{k+1} + \cdots + a_{k+10} > 0.$$

Subtracting the first expression from the second, we obtain

$$a_{k+7} + a_{k+8} + a_{k+9} + a_{k+10} > 0. \tag{1}$$

Hence the sum of any four consecutive terms starting from the eighth term is always positive. Thus,

$$a_8 + a_9 + a_{10} + 2a_{11} + a_{12} + a_{13} + a_{14}$$
$$= (a_8 + a_9 + a_{10} + a_{11}) + (a_{11} + a_{12} + a_{13} + a_{14}) > 0.$$

By assumption, $a_8 + a_9 + a_{10} + a_{11} + a_{12} + a_{13} + a_{14} < 0$, and consequently, $a_{11} > 0$. Similarly, $a_{12} > 0$ and $a_{13} > 0$. Hence $a_{11} + a_{12} + a_{13} > 0$.

Since $a_{11} + a_{12} + a_{13} + \cdots + a_{17} < 0$, it follows that $a_{14} + a_{15} + a_{16} + a_{17} < 0$, which contradicts (1). Thus $n \leq 16$.

Next, we construct a sequence for $n = 16$ satisfying the assumptions. Let

$$S_1 = a_1,$$
$$S_2 = a_1 + a_2,$$
$$\cdots\cdots\cdots\cdots\cdots$$
$$S_{16} = a_1 + a_2 + \cdots + a_{16}.$$

The assumptions may be interpreted as

$$S_{10} < S_3 < S_{14} < S_7 < 0 < S_{11} < S_4 < S_{15} < S_8 < S_1 < S_{12} < S_5$$
$$< S_{16} < S_9 < S_2 < S_{13} < S_6,$$

which is actually equivalent. (Indeed we may also prove that S_{17} does not exist in this way, for otherwise, we obtain $S_6 < S_{17} < S_{10} < S_6$, which is absurd.)

Thus, it is quite easy to write such a sequence $a_1, a_2, \ldots, a_{16}$. For example, put

$$S_{10} = -4, \quad S_3 = -3, \quad S_{14} = -2, \quad S_7 = -1, \quad S_{11} = 1, \quad S_4 = 2,$$

$$S_{15} = 3, \quad S_8 = 4, \quad S_1 = 5, \quad S_{12} = 6, \quad S_5 = 7, \quad S_{16} = 8, \quad S_9 = 9,$$

$$S_2 = 10, \quad S_{13} = 11, \quad S_6 = 12.$$

Then

$$(a_1, a_2, a_3, a_4, \ldots, a_{16}) = (5, 5, -13, 5, 5, 5, -13, 5, 5, -13, 5, 5, 5, -13, 5, 5).$$

Solution 2. Suppose we have a sequence with 17 terms $a_1, a_2, \ldots, a_{17}$ satisfying the assumptions in the problem. We shall use the following identity

$$\sum_{i=1}^{11} (a_i + a_{i+1} + a_{i+2} + \cdots + a_{i+6}) = \sum_{j=1}^{7} (a_j + a_{j+1} + a_{j+2} + \cdots + a_{j+10}).$$

For every $1 \le i \le 11$, it holds $a_i + a_{i+1} + \cdots + a_{i+6} < 0$, and for every $1 \le j \le 7$, we have $a_j + a_{j+1} + \cdots + a_{j+10} > 0$. Thus, the left-hand side is negative and the right-hand side is positive, which is absurd. There are at most 16 terms in this sequence. The construction of an example with 16 terms is the same as in the Solution 1.

【Score Situation】This particular problem saw the following distribution of scores among contestants: 13 contestants scored 6 points, no contestant scored 5 points, 1 contestant scored 4 points, 6 contestants scored 3 points, 2 contestants scored 2 points, 9 contestants scored 1 point, and 6 contestants scored 0 point. The average score of this problem is 3.054, indicating that it was relatively straightforward.

Among the top five teams in the team scores, the United States team achieved a total score of 202 points, the Soviet Union team achieved a total score of 192 points, the Hungary team achieved a total score of 190 points, the United Kingdom team achieved a total score of 190 points, and the Netherlands team achieved a total score of 185 points.

The gold medal cutoff for this IMO was set at 34 points (with 13 contestants earning gold medals), the silver medal cutoff was 24 points (with 29 contestants earning silver medals), and the bronze medal cutoff was 17 points (with 35 contestants earning bronze medals).

In this IMO, a total of five contestants achieved a perfect score of 40 points.

Problem 3.3 (IMO 32-3, proposed by China). Let $S = \{1, 2, 3, \ldots, 280\}$. Find the smallest integer n such that each n-element subset of S contains five numbers which are pairwise relatively prime.

Solution. We first estimate a lower bound of n.

Let $A_i (i = 2, 3, 5, 7)$ be the set consisting of all elements of S which are divisible by i. Denote $A = A_2 \cup A_3 \cup A_5 \cup A_7$. By the inclusion–exclusion principle,

$$
\begin{aligned}
|A| = &|A_2| + |A_3| + |A_5| + |A_7| - |A_2 \cap A_3| - |A_2 \cap A_5| \\
&- |A_2 \cap A_7| - |A_3 \cap A_5| - |A_3 \cap A_7| - |A_5 \cap A_7| \\
&+ |A_2 \cap A_3 \cap A_5| + |A_2 \cap A_3 \cap A_7| + |A_2 \cap A_5 \cap A_7| \\
&+ |A_3 \cap A_5 \cap A_7| - |A_2 \cap A_3 \cap A_5 \cap A_7| \\
= &\left[\frac{280}{2} \right] + \left[\frac{280}{3} \right] + \left[\frac{280}{5} \right] + \left[\frac{280}{7} \right] - \left[\frac{280}{2 \times 3} \right] - \left[\frac{280}{2 \times 5} \right] \\
&- \left[\frac{280}{2 \times 7} \right] - \left[\frac{280}{3 \times 5} \right] - \left[\frac{280}{3 \times 7} \right] - \left[\frac{280}{5 \times 7} \right] + \left[\frac{280}{2 \times 3 \times 5} \right] \\
&+ \left[\frac{280}{2 \times 3 \times 7} \right] + \left[\frac{280}{2 \times 5 \times 7} \right] + \left[\frac{280}{3 \times 5 \times 7} \right] - \left[\frac{280}{2 \times 3 \times 5 \times 7} \right] \\
= &\, 140 + 93 + 56 + 40 - 46 - 28 - 20 - 18 - 13 - 8 + 9 + 6 + 4 + 2 - 1 \\
= &\, 216.
\end{aligned}
$$

If we take the 216-element subset A, then any five numbers of A contain two belonging to the same set A_2, A_3, A_5, or A_7. Therefore, these two are not coprime. Hence, it is impossible to choose from A five numbers that are pairwise coprime. This shows that $n \geq 217$.

Next, we show that there exist five mutually coprime numbers in any 217-element subset of S. Let

$$
\begin{aligned}
B_1 &= \{1 \text{ and all prime numbers in } S\}, \\
B_2 &= \{2^2, 3^2, 5^2, 7^2, 11^2, 13^2\}, \\
B_3 &= \{2 \times 131, 3 \times 89, 5 \times 53, 7 \times 37, 11 \times 23, 13 \times 19\}, \\
B_4 &= \{2 \times 127, 3 \times 83, 5 \times 47, 7 \times 31, 11 \times 19, 13 \times 17\}, \\
B_5 &= \{2 \times 113, 3 \times 79, 5 \times 43, 7 \times 29, 11 \times 17\}, \\
B_6 &= \{2 \times 109, 3 \times 73, 5 \times 41, 7 \times 23, 11 \times 13\}, \\
B &= B_1 \cup B_2 \cup B_3 \cup B_4 \cup B_5 \cup B_6.
\end{aligned}
$$

All numbers in B_i ($i = 1, 2, \ldots, 6$) are mutually coprime, and $B_i \cap B_j = \varnothing$ ($1 \le i < j \le 6$). Since $|B_1| = 60$, $|B_2| = |B_3| = |B_4| = 6$, and $|B_5| = |B_6| = 5$,

$$|B| = 60 + 6 \times 3 + 5 \times 2 = 88, \quad \text{and} \quad |S \backslash B| = 280 - 88 = 192.$$

Any 217 elements of S contain at least $217 - 192 = 25$ elements in B. By the pigeonhole principle, at least $\left[\frac{25}{6}\right] + 1 = 5$ numbers of these 25 elements belong to some B_i ($1 \le i \le 6$), and these five numbers are pairwise coprime.

In conclusion, the smallest n with the required property is 217.

【Score Situation】 This particular problem saw the following distribution of scores among contestants: 23 contestants scored 7 points, 5 contestants scored 6 points, 16 contestants scored 5 points, 37 contestants scored 4 points, 111 contestants scored 3 points, 16 contestants scored 2 points, 14 contestants scored 1 point, and 90 contestants scored 0 point. The average score of this problem is 2.558, indicating that it had a certain level of difficulty.

Among the top five teams in the team scores, the scores of this problem are as follows: the Soviet Union team scored 34 points (with a total team score of 241 points), the China team scored 31 points (with a total team score of 231 points), the Romania team scored 27 points (with a total team score of 225 points), the Germany team scored 33 points (with a total team score of 222 points), and the United States team scored 20 points (with a total team score of 212 points).

The gold medal cutoff for this IMO was set at 39 points (with 20 contestants earning gold medals), the silver medal cutoff was 31 points (with 51 contestants earning silver medals), and the bronze medal cutoff was 19 points (with 84 contestants earning bronze medals).

In this IMO, a total of nine contestants achieved a perfect score of 42 points.

Problem 3.4 (IMO 40-3, proposed by Belarus). Consider an $n \times n$ square board, where n is a fixed even positive integer. The board is divided into n^2 unit squares. We say that two different squares on the board are adjacent if they have a common side.

Mark N unit squares on the board in such a way that every square (marked or unmarked) on the board is adjacent to at least one marked square.

Determine the smallest possible value of N.

Solution. Suppose $n = 2k$. Color the squares black and white like a chessboard. Let $f(n)$ be the smallest possible value of N. Define $f_w(n)$ as the minimum number of white squares that must be marked in order that every black square on the board is adjacent to at least one white-marked square. Define $f_b(n)$ analogously, with the roles of "black" and "white" interchanged. These roles are fully symmetric because n is even. Thus,

$$f_w(n) = f_b(n), \tag{1}$$

$$f(n) = f_w(n) + f_b(n). \tag{2}$$

For convenience, we rotate the board so that the black main diagonal is horizontal. Thus, the numbers of black squares of each black row from top to bottom are $2, 4, \ldots, 2k, \ldots, 4, 2$, as shown in Figure 3.2.

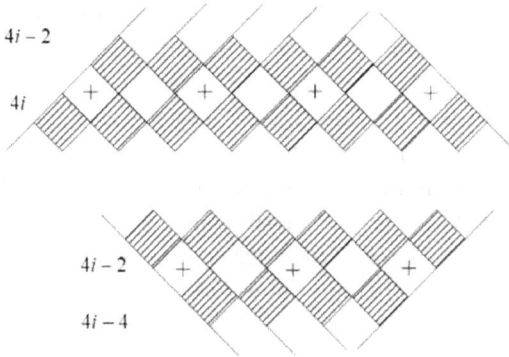

Figure 3.2 The Rotated Board

We mark the white squares only in the white row immediately below the black row with $4i - 2$ black squares, and we mark all odd positioned white squares in these rows. If the white row is above the main diagonal, we mark $2i$ white squares, and we mark $2i - 1$ white squares if it is below the main diagonal (as illustrated in the picture). Thus, the number of marked white squares is

$$2 + 4 + \cdots + k + \cdots + 3 + 1 = \frac{k(k+1)}{2}.$$

It is easy to see that every black square is adjacent to some white-marked square. Hence

$$f_w(n) \leq \frac{k(k+1)}{2}. \tag{3}$$

Observe that any two of the white-marked $\frac{k(k+1)}{2}$ squares have no common adjacent black squares, and thus we need to mark at least $\frac{k(k+1)}{2}$ black squares in order to make each of these white square adjacent to some black-marked square. Consequently,

$$f_b(n) \geq \frac{k(k+1)}{2}. \tag{4}$$

In view of (1), (3), and (4), we obtain

$$f_w(n) = f_b(n) = \frac{k(k+1)}{2}. \tag{5}$$

It follows from (2) and (5) that $f(n) = k(k+1) = \frac{n(n+2)}{4}$.

【Score Situation】This particular problem saw the following distribution of scores among contestants: 23 contestants scored 7 points, 8 contestants scored 6 points, 3 contestants scored 5 points, 20 contestants scored 4 points, 44 contestants scored 3 points, 79 contestants scored 2 points, 119 contestants scored 1 point, and 154 contestants scored 0 point. The average score of this problem is 1.584, indicating that it was relatively challenging.

Among the top five teams in the team scores, the scores of this problem are as follows: the China team scored 11 points (with a total team score of 182 points), the Russia team scored 21 points (with a total team score of 182 points), the Vietnam team scored 9 points (with a total team score of 177 points), the Romania team scored 10 points (with a total team score of 173 points), and the Bulgaria team scored 10 points (with a total team score of 170 points).

The gold medal cutoff for this IMO was set at 28 points (with 38 contestants earning gold medals), the silver medal cutoff was 19 points (with 70 contestants earning silver medals), and the bronze medal cutoff was 12 points (with 118 contestants earning bronze medals).

In this IMO, no contestant achieved a perfect score of 42 points.

3.3　Summary

In the first 64 IMOs, there were a total of four extremal combinatorial problems, as depicted in Figure 3.3. The score details for these problems are presented in Table 3.2. Due to the smaller number of participating teams and missing contestant score information in early IMOs, there are several blanks in Table 3.2.

These four problems were proposed by four countries. The United States, Vietnam, China, and Belarus each contributed one problem.

From Table 3.2, it can be observed that in the first 64 IMOs, there was one extremal combinatorial problem with an average score of 1–2 points; one problem with an average score of 2–3 points; two problems with an average score of 3–4 points. Among these problems, the one with the lowest average score is Problem 3.4 (IMO 40-3), proposed by Belarus.

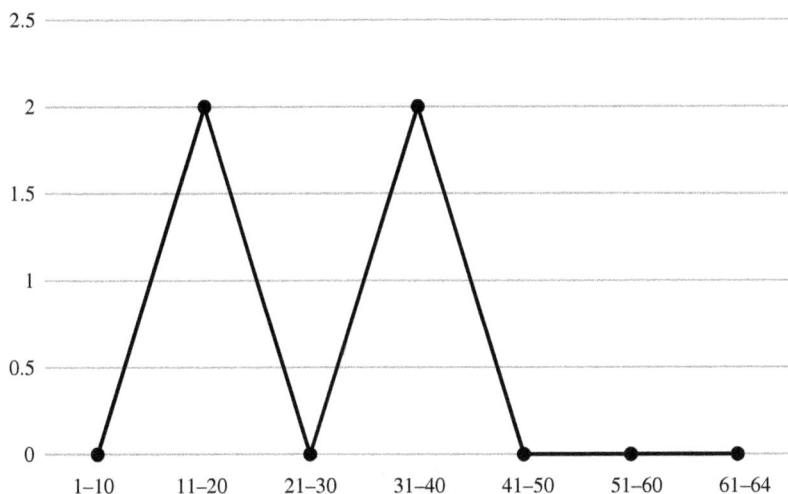

Figure 3.3 Numbers of Extremal Combinatorial Problems in the First 64 IMOs

Table 3.2 Score Details of Extremal Combinatorial Problems in the First 64 IMOs

Problem	3.1	3.2	3.3	3.4
Full points	6	6	7	7
Average score	3.755	3.054	2.558	1.584
Top five mean	4.925		4.833	2.033
6th–15th mean	3.763		3.717	2.800
16th–25th mean			2.800	2.045
Problem number in the IMO	18-4	19-2	32-3	40-3
Proposing country	The United States	Vietnam	China	Belarus

Note. Top five mean = Total score of the top five teams ÷ Total number of contestants from the top five teams,

6th–15th mean = Total score of the 6th–15th teams ÷ Total number of contestants from the 6th–15th teams,

16th–25th mean = Total score of the 16th–25th teams ÷ Total number of contestants from the 16th–25th teams.

Chapter 4

Operation and Logical Reasoning Problems

In recent years, operation problems have become increasingly prevalent in mathematics competitions. Unlike traditional combinatorics problems such as enumerative combinatorics problems, extremum problems, and combinatorial geometry problems, operation problems often appear in novel game formats. They do not require extensive mathematical knowledge but place a greater emphasis on examining students' logical thinking, creativity, and construction skills. Moreover, proofs in solving these problems are unconventional and not readily apparent.

In the first 64 IMOs, there had been a total of 16 operation and logical reasoning problems, accounting for approximately 18.4% of all combinatorics problems. These problems can be primarily categorized into three types: (1) logical reasoning problems, totaling two problems; (2) single-person operation problems, totaling nine problems; (3) double-person operation problems, totaling five problems. The statistical distribution of these three types of problems in the previous IMOs is presented in Table 4.1.

Single-person operation problems typically involve one individual performing actions, with the question of whether a predetermined goal can be achieved after a finite number of moves. Double-person operation problems generally involve two individuals engaging in a game, with the question of which side has a winning strategy. Additionally, two early simple logical reasoning problems are also included in this chapter.

From Table 4.1, it can be observed that operation problems have primarily appeared in the past few decades. These problems are typically positioned as Problem 3 or Problem 6, known for their high difficulty and

Table 4.1 Numbers of Operation and Logical Reasoning Problems in the First 64 IMOs

Content	Session							Total
	1–10	11–20	21–30	31–40	41–50	51–60	61–64	
Logical reasoning problems	2	0	0	0	0	0	0	2
Single-person operation problems	0	0	1	3	2	2	1	9
Double-person operation problems	0	0	0	1	0	3	1	5
Combinatorics problems	7	11	16	16	11	19	7	87
Percentage of operation and logical reasoning problems among combinatorics problems	28.6%	0.0%	6.3%	25.0%	18.2%	26.3%	28.6%	18.4%

low scoring rates. For example, the double-person operation problem (IMO 58-3) had an average score of only 0.042 points, making it the lowest scoring problem in the history of the IMO.

It is important to note that for each problem, the solutions are followed by information on the scores, including the number of contestants in each score range, the average score, and the scores of the top five teams. However, early IMOs often lacked information on contestant scores, so the number of contestants in each score range only represents the counted number of contestants, and some problems lack scores of the top five teams.

4.1 Common Theorems, Formulas, and Methods

4.1.1 *Common methods for single-person operation problems*

(1) *Method of adjustment*

Example 4.1. Define the following operation for a positive integer n: If n is even, divide n by 2 and obtain $\frac{n}{2}$; if n is odd, first add 1 to n and then divide the sum by 2. Prove that the number 1 will appear after a finite number of operations starting from any positive integer n.

Proof. For any positive integer n, let $f = n$.

If $f \neq 1$, by definition, we obtain f_1 after an operation given by the following formula

$$f_1 = \begin{cases} \dfrac{n}{2} & (n \text{ is even}), \\ \dfrac{n+1}{2} & (n \geq 3 \text{ is odd}), \end{cases}$$

then

$$f - f_1 = \begin{cases} \dfrac{n}{2} & (n \text{ is even}), \\ \dfrac{n-1}{2} & (n \geq 3 \text{ is odd}), \end{cases} \qquad f - f_1 \geq 1.$$

Thus if $f \neq 1$, then the value strictly decreases after an operation. Since it is impossible to have an infinite descent in positive integers, 1 will appear sooner or later.

(2) Method of invariants

Example 4.2. Three integers are written on a blackboard. An operation consists of erasing one number and then replacing it with $a + b - 1$, where a and b are the remaining two numbers on the blackboard. Assume that after several operations, we get $(17, 2011, 2027)$ on the board. Is it possible that the initial numbers on the board are $(2, 2, 2)$ or $(3, 3, 3)$?

Solution. The numbers $(2, 2, 2)$ are impossible. If initially the numbers on the board were $(2, 2, 2)$, then the parities after each operation would be

$$(\text{even, even, even}) \to (\text{even, even, odd}) \to (\text{even, even, odd})$$

(this parity pattern is preserved after each operation), so it cannot reach $(17, 2011, 2027)$.

We claim the numbers $(3, 3, 3)$ are possible.

Note that from $(17, a, a + 16)$, we can get $(17, a + 16, a + 32)$. An easy induction shows that we can get $(17, a + 16k, a + 16(k + 1))$ for any nonnegative integer k. Since $(17, 2011, 2027) = (17, 11 + 16 \times 125, 11 + 16 \times 126)$, it suffices to obtain $(17, 11, 27)$. By the reversal operation, we see that $(17, 11, 27)$ is available from $(3, 3, 3)$;

$$(17, 11, 27) \leftarrow (7, 11, 17) \leftarrow (5, 7, 11) \leftarrow (3, 5, 7) \leftarrow (3, 3, 5) \leftarrow (3, 3, 3).$$

(3) *Mathematical induction*

Example 4.3. There are 64 unit cubes, each of which has a red face and 5 white faces. Place these 64 cubes on an 8×8 chess board (each square has side length 1 and exactly one cube is placed on each square), and then rotate the wooden blocks. The rotation rule is to rotate eight cubes in the same row (or column) simultaneously in the same direction. Can all the red faces of the cubes be turned up after a finite number of rotations?

Solution. We shall prove a more general result: there are n^2 cubes placing on an $n \times n$ board, and the answer is yes. We induct on n.

For $n = 1$, the statement is obvious.

Assume that the statement is true for $n = k$. Consider the case $n = k + 1$. By the inductive hypothesis, we can turn up all the red faces of the upper left $k \times k$ cubes after a finite number of operations. Now rotate the first column so that the red faces of the first k cubes are facing leftward.

Using the inductive hypothesis, we can turn up the red faces of the upper right $k \times k$ cubes, which does not change the position of the red faces of the cubes on the first column. Rotate the first column again so that the red faces of the first k cubes are facing upward. Now we have all red faces up in the upper $k \times (k + 1)$ squares.

Now rotate the first row $90°$. A similar strategy as above can turn up all the red faces in the lower $k \times (k + 1)$ squares, which does not change the position of the red faces of the cubes in the first row. Finally, we rotate the first row to turn up all the red faces. This proves the statement for $n = k + 1$. By induction, the statement is true for all positive integers n. And in particular, it is true for $n = 8$.

(4) *Reversal argument*

Example 4.4. There are 49 number of 1's and 50 number of 0's arranged on a circle. We do the following operation: write a 0 between two equal numbers and a 1 between two different numbers, and then erase all the original numbers. Is it possible to reach the state that all numbers on the circle are 0?

Solution. Suppose we obtain the state of all 0's after the kth operation for the first time. Then after the $(k - 1)$th operation, there are all 1's on the circle; after the $(k - 2)$th operation, 0 and 1 are arranged alternatively on the circle. However, there are totally 99 numbers, which is odd. Hence, it is impossible. Thus, it is impossible to have all 0's on the circle.

(5) *Proof by contradiction*

Example 4.5. Five nonzero rational numbers a, b, c, d, and e are arranged on a circle. Each operation replaces the five numbers by the differences of neighbouring numbers, that is we write $|a - b|$, $|b - a|$, $|c - d|$, $|d - e|$, and $|e - a|$ on the circle in this order. Prove that 0 must appear after a finite number of operations.

Proof. Assume on the contrary that 0 never appears. First consider the case that a, b, c, d, and e are integers. After the 1st move, the five numbers will remain positive integers. For $x, y > 0$, clearly $|x - y| < \max\{x, y\}$, so the maximum number will decrease (by at least 1) after each operation. However, it is impossible to have an infinite descent in positive integers, from which we reach a contradiction. For rational numbers, we may multiply the least common multiple of their denominators to turn the five numbers into integers and reduce the problem to the case of integers.

4.1.2 *Common methods for double-person operation problems*

(1) *Recursive method*

Some double-person operation problems can be deduced from the final state to determine the outcome of any initial state.

Example 4.6. A box contains 200 matches. Two players A and B alternatively take matches from the box, taking at least 1 match each time and up to 20 matches at most. Player A goes first. The player who receives the last match wins. Who has a winning strategy?

Solution. Since $200 = 21 \times 9 + 11$ is not a multiple of 21, player A has a winning strategy. Player A takes 11 matches in his first turn. After that, whenever player B takes $k(1 \leq k \leq 20)$ matches, player A takes $21 - k$ matches immediately. Thus, after player A's turn, the number of matches in the box remains a multiple of 21; and after player B's turn, the number of matches in the box is not divisible by 21. Therefore, player A will take the last match and win the game.

(2) *Correspondence method*

If one player can guarantee that he will always have a step to take, then he will never lose, and in a two-player game without a draw, it is equivalent to winning.

Example 4.7. Two players A and B play the following game: A and B alternatively write a positive integer $\leq P$ on a blackboard and the written number cannot be a divisor of some previously written number. Player A goes first. The player who cannot write a new number loses.

(a) Who has a winning strategy for $P = 10$?
(b) Who has a winning strategy for $P = 1000$?

Solution. (a) Player A has a winning strategy. He first writes 6, and then 1, 2, and 3 are forbidden. Divide the remaining six numbers into three groups: (4, 5), (7, 9), and (8, 10). In the following moves, player A can always write the number in the same group as what player B previously wrote. Thus, player A has a winning strategy.

(b) Consider a new game with the same rule but the numbers are limited to $[2, 1000]$ with the same rule. If player A has a winning strategy in this new game, then he also has a winning strategy for the original game since in the new game, the number 1 is forbidden after his first move. If player B has a winning strategy for the new game, then player A can write 1 first and let player B go first in the new game. Thus, player A also has a winning strategy in the original game. Therefore, player A has a winning strategy in any case.

(3) *Estimation of extremal values*

In some two-player game with scores, if player A has a strategy to score at least a points and player B has a strategy to prevent player A from scoring more than a points, then a is the optimum value that player A can guarantee to get.

Example 4.8. Two players A and B place plus and minus signs before $1, 2, \ldots, 20$ alternatively with A goes first. After all 20 signs are determined, let S be the absolute value of the algebraic sum of these signed numbers. Player A aims to minimize S, however player B aims to maximize it. What is the maximum value of S that player B can guarantee?

Solution. Strategy of player B: divide the numbers into 10 pairs: $(1, 2), (3, 4), \ldots, (19, 20)$. Whenever player A determines a sign of some number, player B determines the sign of the number in the same groups. Player B can assure that 19 and 20 have the same sign, and the other pairs have the different sign. In this way, player B can assure that $S \geq 20 + 19 - 9 \times 1 = 30$.

Strategy for player A: Naturally player A wants to cancel out. Let S_i be the algebraic sum of the $2i$ signed numbers after the ith move of B. Player A chooses the largest remaining number a_{i+1} and places the opposite sign as S_i (if $S_i = 0$, then player A can place any sign). Assume that the last time player A changes the sign of the sum is after his kth move ($1 \leq k \leq 10$), i.e., $a_k \geq |S_{k-1}|$. Then

$$|S_{k-1} \pm a_k| \leq a_k \leq 21 - k$$

since $20 = a_1 > a_2 > \cdots > a_{k-1} > a_k$ and $a_k \leq 20 - (k-1) = 21 - k$.

After the kth move of B, he places a sign before $b_k \leq a_k - 1 \leq 20 - k$. Thus

$$|S_k| = |S_{k-1} \pm a_k \pm b_k| \leq |S_{k-1} \pm a_k| + b_k \leq 41 - 2k.$$

Afterward, we have

$$b_j < a_j < |S_{j-1}| \quad \text{for } j > k.$$

Hence

$$|S_j| \leq |S_{j-1}| - |a_j - b_j| \leq |S_{j-1}| - 1.$$

Consequently,

$$S = |S_{10}| = |S_k| + \sum_{j=k+1}^{10} (|S_j| - |S_{j-1}|) \leq 41 - 2k - (10 - k)$$

$$= 31 - k \leq 30.$$

In view of the strategies of both players, the maximum value of S that player B can guarantee is 30.

4.2 Problems and Solutions

4.2.1 *Logical reasoning problems*

Problem 4.1 (IMO 5-6, proposed by Hungary). Five students, A, B, C, D, and E, took part in a contest. One prediction was that the contestants would finish in the order $ABCDE$. This prediction was very poor. In fact, no contestant finished in the position predicted, and no two contestants predicted to finish consecutively actually did so. A second prediction had the contestants finishing in the order $DAECB$. This prediction was better. Exactly two of the contestants finished in the places predicted, and two disjoint pairs of students predicted to finish consecutively actually did so.

Determine the order in which the contestants finished.

Solution. The two predictions are listed in the following Table 4.2.

Table 4.2 Two Predictions

Order	1	2	3	4	5
First prediction	A	B	C	D	E
Second prediction	D	A	E	C	B

Since the second prediction was correct on two disjoint pairs of consecutive students, two contestants of such a pair are either both in the correct place or both in the incorrect place. Since exactly two contestants finished in the correct place as the second prediction, one of the two consecutive pairs is in the correct place and the other pair is not.

We have three possible combinations for two pairs of consecutive contestants for the second prediction, namely (DA, EC), (DA, CB), and (AE, CB). Only for (DA, CB) can we move one consecutive pair to another consecutive place.

If DA position is in the correct place, then C and B are not, so the correct order is $DACBE$; however, E is in the correct place for the first prediction, a contradiction.

If CB is in the correct place, then D and A are not, and thus the correct order is $EDACB$, which is indeed the solution.

【Score Situation】This particular problem saw the following distribution of scores among contestants: 2 contestants scored 8 points, 6 contestants scored 7 points, 4 contestants scored 6 points, 1 contestant scored 5 points, 1 contestant scored 4 points, 2 contestants scored 3 points, no contestant scored 2 points, no contestant scored 1 point, and no contestant scored 0 point. The average score of this problem is 6.063, indicating that it was simple.

Among the top five teams in the team scores, the Soviet Union team achieved a total score of 271 points, the Hungary team achieved a total score of 234 points, the Romania team achieved a total score of 191 points, the Yugoslavia team achieved a total score of 162 points, and the Czechoslovakia team achieved a total score of 151 points.

The gold medal cutoff for this IMO was set at 35 points (with 7 contestants earning gold medals), the silver medal cutoff was 28 points (with 11 contestants earning silver medals), and the bronze medal cutoff was 21 points (with 17 contestants earning bronze medals).

In this IMO, no contestant achieved a perfect score of 40 points.

Problem 4.2 (IMO 16-1, proposed by the United States). Three players A, B, and C play the following game: On each of three cards an integer is written. These three numbers p, q, and r satisfy $0 < p < q < r$. The three cards are shuffled, and one is dealt to each player. Each then receives the number of counters indicated by the card he holds. Then the cards are shuffled again; the counters remain with the players.

This process (shuffling, dealing, and giving out counters) takes place for at least two rounds. After the last round, A has 20 counters in all, B has 10, and C has 9. At the last round B received r counters. Who received q counters in the first round?

Solution. Let n be the number of rounds played. Then

$$n(p + q + r) = 20 + 10 + 9 = 39 = 3 \times 13.$$

By assumption, $n \geq 2$ and $p + q + r \geq 1 + 2 + 3 = 6$. We conclude that $n = 3$, and $p + q + r = 13$.

Since B received 10 counters in all, and he received at least two counters in the first two rounds, we see that $r \leq 10 - 2 = 8$.

Since the number of counters received by A does not exceed $2r + q < 3r$, hence $3r > 20$, i.e., $r \geq 7$, and consequently, $r = 7$ or 8.

If $r = 7$, then $p + q = 6$, and B received three counters in the first two rounds. It is impossible to have two numbers from p, q, r adding up to 3.

If $r = 8$, then $p + q = 5$, and B received two counters in the first two rounds. Thus, $p = 1$ and $q = 4$, and B received $p = 1$ card in each of the first two rounds.

As for A, it is only possible to write $20 = 8 + 8 + 4$. Player A received eight counters in each of the first two rounds and 4 counters in the final round. Thus, it is C who received $q = 4$ counters in the first round.

【Score Situation】 This particular problem saw the following distribution of scores among contestants: 113 contestants scored 5 points, 9 contestants scored 4 points, 6 contestants scored 3 points, 6 contestants scored 2 points, 1 contestant scored 1 point, and 5 contestants scored 0 point. The average score of this problem is 4.514, indicating that it was simple.

Among the top five teams in the team scores, the scores of this problem are as follows: the Soviet Union team scored 40 points (with a total team score of 256 points), the United States team scored 40 points (with a total team score of 243 points), the Hungary team scored 40 points (with a total team score of 237 points), the German Democratic Republic team scored 40 points (with a total team score of 236 points), and the Yugoslavia team scored 40 points (with a total team score of 216 points).

The gold medal cutoff for this IMO was set at 38 points (with 10 contestants earning gold medals), the silver medal cutoff was 30 points (with 24 contestants earning silver medals), and the bronze medal cutoff was 23 points (with 37 contestants earning bronze medals).

In this IMO, a total of six contestants achieved a perfect score of 40 points.

4.2.2 *Single-person operation problems*

Problem 4.3 (IMO 27-3, proposed by the German Democratic Republic). To each vertex of a regular pentagon an integer is assigned in such a way that the sum of all five numbers is positive. If three consecutive vertices are assigned the numbers x, y, and z, respectively and $y < 0$ then the following operation is allowed: the numbers x, y, and z are replaced by $x+y$, $-y$, and $z+y$, respectively. Such an operation is performed repeatedly as long as at least one of the five numbers is negative.

Determine whether this procedure necessarily comes to an end after a finite number of steps.

Solution 1. The answer is affirmative.

Denote the five numbers in order by x, y, z, u, v, where v and x are adjacent. Assume that $y < 0$. Let S be the sum of squares of the five numbers and squares of each sum of two adjacent numbers. After the operation which replaces x, y, z with $x + y, -y, z + y$, the difference of S is given by

$$
\begin{aligned}
(x + y)^2 &+ (-y)^2 + (z + y)^2 + u^2 + v^2 + x^2 + z^2 + (z + y + u)^2 \\
&+ (u + v)^2 + (v + x + y)^2 - [x^2 + y^2 + z^2 + u^2 + v^2 + (x + y)^2 \\
&+ (y + z)^2 + (z + u)^2 + (u + v)^2 + (v + x)^2] \\
&= 2y(x + y + z + u + v) < 0.
\end{aligned}
$$

Thus, S strictly decreases after each operation. Since S attains only nonnegative integers, this procedure necessarily terminates after a finite number of steps.

Solution 2. Denote the five numbers in order by u_1, u_2, u_3, u_4, u_5, where u_5 and u_1 are adjacent. Define

$$
f(u_1, u_2, \ldots, u_5) = \sum_{i=1}^{5} |u_i| + \sum_{i=1}^{5} |u_i + u_{i+1}| + \sum_{i=1}^{5} |u_i + u_{i+1} + u_{i+2}|
$$

$$
+ \sum_{i=1}^{5} |u_i + u_{i+1} + u_{i+2} + u_{i+3}|,
$$

where $u_{5+i} = u_i$ for $i = 1, 2, 3$.

Assume without loss of generality that $u_2 < 0$. Then the five numbers become v_1, v_2, v_3, v_4, and v_5 after the operation, where $v_1 = u_1 + u_2$, $v_2 = -u_2$, $v_3 = u_3 + u_2$, $v_4 = u_4$, and $v_5 = u_5$. Hence

$$f(v_1, v_2, v_3, v_4, v_5) - f(u_1, u_2, u_3, u_4, u_5)$$
$$= |u_1 + u_3 + u_4 + u_5 + 2u_2| - |u_1 + u_3 + u_4 + u_5|.$$

Since $u_1 + u_2 + u_3 + u_4 + u_5 > 0$ and $u_2 < 0$, we have

$$-(u_1 + u_3 + u_4 + u_5) < u_1 + u_3 + u_4 + u_5 + 2u_2 < u_1 + u_3 + u_4 + u_5.$$

Hence, $|u_1 + u_3 + u_4 + u_5 + 2u_2| < u_1 + u_3 + u_4 + u_5$ and

$$f(v_1, v_2, v_3, v_4, v_5) < f(u_1, u_2, u_3, u_4, u_5).$$

Thus, f strictly decreases after each operation. Since f attains only nonnegative integers, this procedure necessarily terminates after a finite number of steps.

【Score Situation】 This particular problem saw the following distribution of scores among contestants: 12 contestants scored 7 points, 1 contestant scored 6 points, 3 contestants scored 5 points, 5 contestants scored 4 points, 2 contestants scored 3 points, 10 contestants scored 2 points, 32 contestants scored 1 point, and 145 contestants scored 0 point. The average score of this problem is 0.871, indicating that it was extremely difficult.

Among the top five teams in the team scores, the scores of this problem are as follows: the Soviet Union team scored 22 points (with a total team score of 203 points), the United States team scored 17 points (with a total team score of 203 points), the Germany team scored 10 points (with a total team score of 196 points), the China team scored 14 points (with a total team score of 177 points), and the German Democratic Republic team scored 12 points (with a total team score of 172 points).

The gold medal cutoff for this IMO was set at 34 points (with 18 contestants earning gold medals), the silver medal cutoff was 26 points (with 41 contestants earning silver medals), and the bronze medal cutoff was 17 points (with 48 contestants earning bronze medals).

In this IMO, only three contestants achieved a perfect score of 42 points, namely Vladimir Roganov and Stanislav Smirnov from the Soviet Union, and Géza Kós from Hungary.

Problem 4.4 (IMO 34-3, proposed by Finland). On an infinite chessboard, a game is played as follows. At the start, n^2 pieces are arranged on the chessboard in an n by n block of adjoining squares, one piece in each square. A move in the game is a jump in a horizontal or vertical direction over an adjacent occupied square to an unoccupied square immediately beyond. The piece which has been jumped over is removed.

Find those values of n for which the game can end with only one piece remaining on the board.

Solution. We first show that the game cannot end with only one piece for $3 \mid n$.

Label the squares with A, B, and C as shown below. Thus, the squares and the pieces are divided into three classes:

$\ddots$	$\vdots$	$\vdots$	$\vdots$	$\vdots$	$\vdots$	$\vdots$	$\vdots$	$\vdots$	$\vdots$	$\vdots$	$\vdots$	$\vdots$	$\vdots$	$\iddots$
$\cdots$	A	B	C	A	B	C	A	B	C	A	B	C	A	$\cdots$
$\cdots$	B	C	A	B	C	A	B	C	A	B	C	A	B	$\cdots$
$\cdots$	C	A	B	C	A	B	C	A	B	C	A	B	C	$\cdots$
$\cdots$	A	B	C	A	B	C	A	B	C	A	B	C	A	$\cdots$
$\cdots$	B	C	A	B	C	A	B	C	A	B	C	A	B	$\cdots$
$\cdots$	C	A	B	C	A	B	C	A	B	C	A	B	C	$\cdots$
$\iddots$	$\vdots$	$\vdots$	$\vdots$	$\vdots$	$\vdots$	$\vdots$	$\vdots$	$\vdots$	$\vdots$	$\vdots$	$\vdots$	$\vdots$	$\vdots$	$\ddots$

Each 3×3 square contains 3 A's, 3 B's, and 3 C's. After the ith move, denote the number of pieces in squares labeled A, B, and C by S_A^i, S_B^i, and S_C^i, respectively. For $3 \mid n$, we have

$$S_A^0 = S_B^0 = S_C^0 = \frac{n^2}{3}.$$

Initially, S_A^0, S_B^0, and S_C^0 have the same parity. After each move, the number of pieces in two classes decreases by one and the number of pieces in the third class increases by one. Thus S_A^i, S_B^i, and S_C^i remain the same parity. Consequently, it cannot happen that two classes contain no piece, and one class contains exactly one piece.

For $3 \mid n$, the game cannot end up with only one piece remaining on the board.

Next, we show that it is possible to do so for all $n = 3k + 1$ or $3k + 2$.

First observe that if there are four pieces in L-shaped positions as shown in Figure 4.1, and the three squares under the bottom line are empty, then we may move in the following way as shown in the diagram:

The effect is the same as removing the bottom three pieces directly with the fourth piece remaining unchanged.

We now prove by induction on $n = 3k + 1$ and $3k + 2$. For $k = 0$, there is nothing to prove for $n = 1$. For $n = 2$, we may move in the following way as shown in Figure 4.2.

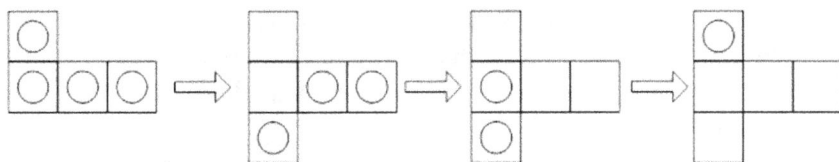

Figure 4.1

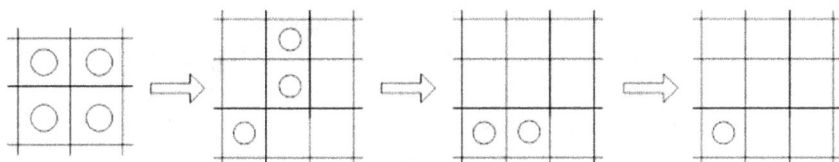

Figure 4.2

Assume that the assertion is true for k. Consider the case of $k + 1$. We divide the initial $n \times n$ array into four parts (as shown in Figure 4.3).

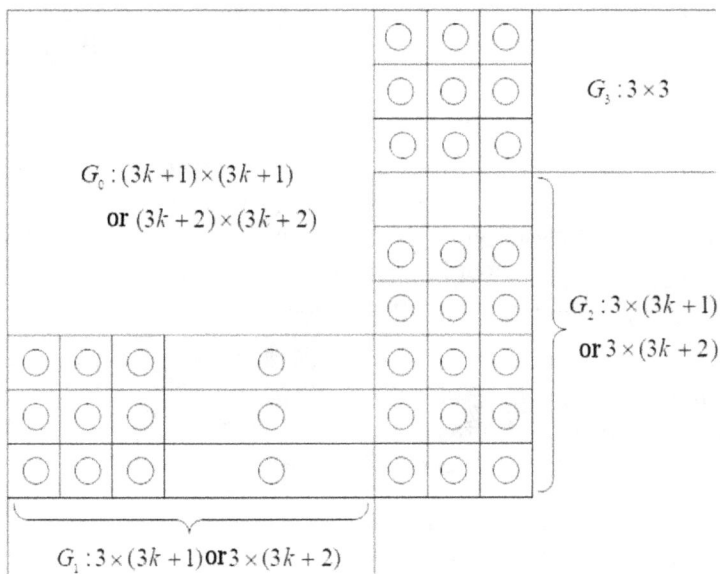

Figure 4.3

By our previous observation, we can remove the pieces in G_1, 3 at a time and column by column from left to right. Then remove the pieces in

G_2, 3 at a time and row by row from bottom to top. Finally, we move the pieces in G_3, 3 at a time and column by column from right to left. Now we are left with the pieces in G_0. By the inductive hypothesis, it is possible to end with only one piece.

Thus, the answer is all positive integers n not divisible by 3.

【Score Situation】 This particular problem saw the following distribution of scores among contestants: 30 contestants scored 7 points, 8 contestants scored 6 points, 7 contestants scored 5 points, 21 contestants scored 4 points, 15 contestants scored 3 points, 6 contestants scored 2 points, 33 contestants scored 1 point, and 293 contestants scored 0 point. The average score of this problem is 1.131, indicating that it was relatively challenging.

Among the top five teams in the team scores, the scores of this problem are as follows: The China team scored 34 points (with a total team score of 215 points), the Germany team scored 24 points (with a total team score of 189 points), the Bulgaria team scored 23 points (with a total team score of 178 points), the Russia team scored 25 points (with a total team score of 177 points), and the Chinese Taiwan team scored 17 points (with a total team score of 162 points).

The gold medal cutoff for this IMO was set at 30 points (with 35 contestants earning gold medals), the silver medal cutoff was 20 points (with 66 contestants earning silver medals), and the bronze medal cutoff was 11 points (with 97 contestants earning bronze medals).

In this IMO, only two contestants achieved a perfect score of 42 points, namely Hong Zhou from China and Hung-Wu Wu from Chinese Taiwan.

Problem 4.5 (IMO 34-6, proposed by the Netherlands).

There are n lamps $L_0, L_1, \ldots, L_{n-1}$ in a circle ($n > 1$), where we denote $L_{n+k} = L_k$. (A lamp always is either on or off.) Perform steps $s_0, s_1, \ldots$ as follows: at step s_i, if L_{i-1} is lit, switch L_i from on to off or vice versa; otherwise do nothing. Initially all the lamps are on. Show that:

(a) There is a positive integer $M(n)$ such that after $M(n)$ steps all the lamps are on again.
(b) If $n = 2^k$, then we can take $M(n) = n^2 - 1$.
(c) If $n = 2^k + 1$, then we can take $M(n) = n^2 - n + 1$.

Proof. (a) Define a 0–1 sequence $x_0, x_1, x_2, \ldots$ as follows: $x_0 = x_1 = \cdots = x_{n-1} = 1$, and for $j \geq n$,

$$x_j \equiv x_{j-1} + x_{j-n} \pmod{2}.$$

If 1 represents "ON" and 0 represents "OFF," then $x_0, x_1, \ldots, x_{n-1}$ correspond to the initial states of the lamps $L_0, L_1, \ldots, L_{n-1}$. Since $x_{j+n} \equiv x_{j+n-1} + x_j \pmod 2$ for $j \geq 0$, an easy induction shows that x_{j+n}

represents the state of lamp $L_j = L_{j+n}$ after step s_j. Thus, after step s_j, the states of the lamps $L_{j+1}, L_{j+2}, \ldots, L_{j+n}$ are given by the n-tuple $(x_{j+1}, x_{j+2}, \ldots, x_{j+n})$. There are 2^n possible n-tuples of 0 and 1, so there exists $0 \le j < l$ such that

$$(x_j, x_{j+1}, \ldots, x_{j+n-1}) = (x_l, x_{l+1}, \ldots, x_{l+n-1}).$$

Now if $j > 0$, then by $x_{j-1} \equiv x_{j+n-1} + x_{j+n-2} \equiv x_{l+n-1} + x_{l+n-2} \equiv x_{l-1} \pmod 2$, we see that

$$(x_{j-1}, x_j, \ldots, x_{j+n-2}) = (x_{l-1}, x_l, \ldots, x_{l+n-2}).$$

Continuing this backward argument, we have

$$(x_0, x_1, \ldots, x_{n-1}) = (x_{l-j}, x_{l-j+1}, \ldots, x_{l-j+n-1}).$$

Thus, after $M(n) = l - j$ steps, all the lamps are on again.

Next we will prove (b) and (c). For brevity, we will omit "mod 2" since all the congruences are modulo 2.

For $j \ge n$, we have $x_j \equiv x_{j-1} + x_{j-n}$. Thus, for $j \ge 2n$,

$$x_j \equiv x_{j-1} + x_{j-n} \equiv (x_{j-2} + x_{j-1-n}) + (x_{j-1-n} + x_{j-2n})$$

$$= x_{j-2} + 2x_{j-1-n} + x_{j-2n}.$$

Assume for some positive integer r, the following holds for $j \ge rn$:

$$x_j \equiv \sum_{i=0}^{r} C_r^i x_{j-(r-i)n-i}. \tag{1}$$

Then for $j \ge (r+1)n$, we have

$$x_j \equiv \sum_{i=0}^{r} C_r^i x_{j-(r-i)n-i} \equiv \sum_{i=0}^{r} C_r^i \left(x_{j-(r-i)n-i-1} + x_{j-(r+1-i)n-i} \right)$$

$$= \sum_{i=0}^{r+1} C_{r+1}^i x_{j-(r+1-i)n-i}.$$

By induction, we see that (1) holds for any positive integer r and $j \ge rn$. For $n = 2^k$, take $r = 2^k = n$, and Kummer's theorem shows that C_r^i is even for $0 < i < r$. Thus $x_j \equiv x_{j-n^2} + x_{j-n}$ for $j \ge rn$. Hence

$$x_{j-n^2} \equiv x_j - x_{j-n} \equiv x_{j-1}.$$

As in the proof of part (a), we see that $\{x_m\}$ has period $n^2 - 1$.

For $n = 2^k + 1$, take $r = 2^k = n - 1$. For $j \geq rn$, we have $x_j \equiv x_{j-(n-1)n} + x_{j-n+1}$. Thus

$$x_{j-n^2+n} \equiv x_j + x_{j-n+1} \equiv x_{j+1}.$$

Therefore, $\{x_m\}$ has period $n^2 - n + 1$.

【Score Situation】This particular problem saw the following distribution of scores among contestants: 34 contestants scored 7 points, 6 contestants scored 6 points, 26 contestants scored 5 points, 20 contestants scored 4 points, 22 contestants scored 3 points, 81 contestants scored 2 points, 38 contestants scored 1 point, and 186 contestants scored 0 point. The average score of this problem is 1.816, indicating that it was relatively challenging.

Among the top five teams in the team scores, the scores of this problem are as follows: The China team scored 26 points (with a total team score of 215 points), the Germany team scored 42 points (with a total team score of 189 points), the Bulgaria team scored 28 points (with a total team score of 178 points), the Russia team scored 33 points (with a total team score of 177 points), and the Chinese Taiwan team scored 25 points (with a total team score of 162 points).

The gold medal cutoff for this IMO was set at 30 points (with 35 contestants earning gold medals), the silver medal cutoff was 20 points (with 66 contestants earning silver medals), and the bronze medal cutoff was 11 points (with 97 contestants earning bronze medals).

In this IMO, only two contestants achieved a perfect score of 42 points, namely Hong Zhou from China and Hung-Wu Wu from Chinese Taiwan.

Problem 4.6 (IMO 37-1, proposed by Finland). We are given a positive integer r and a rectangular board $ABCD$ with dimensions $|AB| = 20$ and $|BC| = 12$. The rectangle is divided into a grid of 20×12 unit squares. The following moves are permitted on the board: one can move from one square to another only if the distance between the centers of the two squares is $\sqrt{r}$. The task is to find a sequence of moves leading from the square with A as a vertex to the square with B as a vertex.

(a) Show that the task cannot be done if r is divisible by 2 or 3.
(b) Prove that the task is possible when $r = 73$.
(c) Can the task be done when $r = 97$?

Proof. We label each square by its row number and column number. Denote by (i, j) the square on the ith row and jth column ($i = 1, 2, \ldots, 12$; $j = 1, 2, \ldots, 20$). Assume the square with A as a vertex is $(1, 1)$, the square with B as a vertex is $(1, 20)$. By the hypothesis, one can move from (i_1, j_1)

to (i_2, j_2) if and only if

$$(i_1 - i_2)^2 + (j_1 - j_2)^2 = r.$$

(a) If $2 \mid r$, then $i_1 - i_2$ and $j_1 - j_2$ have the same parity, i.e.,

$$i_1 - i_2 \equiv j_1 - j_2 \pmod{2},$$

from which $i_1 - j_1 \equiv i_2 - j_2 \pmod 2$. However, $1 - 1 \not\equiv 1 - 20 \pmod 2$, so it is impossible to make a sequence of moves from $(1, 1)$ to $(1, 20)$.

If $3 \mid r$, then the congruence equation

$$(i_1 - i_2)^2 + (j_1 - j_2)^2 \equiv 0 \pmod 3$$

implies

$$i_1 - i_2 \equiv j_1 - j_2 \equiv 0 \pmod 3,$$

because $x^2 \equiv 1 \pmod 3$ if $x \not\equiv 0 \pmod 3$. Thus $i_1 - j_1 \equiv i_2 - j_2 \pmod 3$. However,

$$1 - 1 \not\equiv 1 - 20 \pmod 3,$$

and again it is impossible to make a sequence of moves from $(1,1)$ to $(1,20)$.

(b) If $r = 73$, then $(i_1 - i_2)^2 + (j_1 - j_2)^2 = 73$ holds for $\{|i_1 - i_2|, |j_1 - j_2|\} = \{3, 8\}$. The following moves achieve our task:

$$(1, 1) \to (4, 9) \to (7, 17) \to (10, 9) \to (2, 6) \to (5, 14)$$

$$\to (8, 6) \to (11, 14) \to (3, 17) \to (6, 9) \to (9, 17) \to (1, 20).$$

(c) If $r = 97$, then the equality

$$(i_1 - i_2)^2 + (j_1 - j_2)^2 = 97$$

holds only for $\{|i_1 - i_2|, |j_1 - j_2|\} = \{4, 9\}$. The task cannot be done because this board is too small. To see this, we color the board in the following way, as shown in Figure 4.4. We see that any move from a black square can only have its destination of a black square. Since $(1, 1)$ is black and $(1, 20)$ is white, the task cannot be done.

【Score Situation】This particular problem saw the following distribution of scores among contestants: 63 contestants scored 7 points, 12 contestants scored 6 points, 17 contestants scored 5 points, 111 contestants scored 4 points, 26 contestants scored 3 points, 98 contestants scored 2 points, 30 contestants scored 1 point, and 67 contestants scored 0 point. The average score of this problem is 3.175, indicating that it was relatively straightforward.

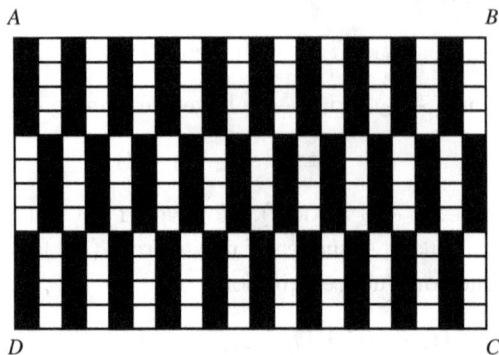

Figure 4.4

Among the top five teams in the team scores, the scores of this problem are as follows: The Romania team scored 23 points (with a total team score of 187 points), the United States team scored 39 points (with a total team score of 185 points), the Hungary team scored 25 points (with a total team score of 167 points), the Russia team scored 28 points (with a total team score of 162 points), and the United Kingdom team scored 39 points (with a total team score of 161 points).

The gold medal cutoff for this IMO was set at 28 points (with 35 contestants earning gold medals), the silver medal cutoff was 20 points (with 66 contestants earning silver medals), and the bronze medal cutoff was 12 points (with 99 contestants earning bronze medals).

In this IMO, only one contestant achieved a perfect score of 42 points, namely Ciprian Manolescu from Romania.

Problem 4.7 (IMO 41-3, proposed by Belarus). Let $n \geq 2$ be a positive integer and λ a positive real number. Initially, there are n fleas on a horizontal line, not all at the same point. We define a move as choosing two fleas, at some point A and B, with A to the left of B, and letting the flea from A jump to the point C to the right of B such that $\frac{BC}{AB} = \lambda$.

Determine all values of λ such that, for any point M on the line and any initial position of the n fleas, there exists a sequence of moves that will take them all to positions right of M.

Solution. The plausibly beneficial strategy to take the fleas far to the right is to choose, in each move, the leftmost flea and let it jump over the rightmost one. With this strategy, we arrive after k moves to a configuration in which two parameters will be of importance: the maximum distance between fleas, which we denote by d_k, and the minimum distance between neighboring fleas, which we denote by δ_k. Clearly,

$$d_k \geq (n-1)\delta_k.$$

After the $(k+1)$th move, a new distance between neighboring fleas appears, namely λd_k. It can be the new minimum distance, so that $\delta_{k+1} = \lambda d_k$; and if not, then certainly $\delta_{k+1} \geq \delta_k$. In any case,

$$\frac{\delta_{k+1}}{\delta_k} \geq \min\left\{1, \frac{\lambda d_k}{\delta_k}\right\} \geq \min\left\{1, (n-1)\lambda\right\}.$$

Thus if $\lambda \geq \frac{1}{n-1}$, then $\delta_{k+1} \geq \delta_k$ for all k; the minimum distance does not decrease. So, the position of the leftmost flea keeps on shifting by steps of size not less than a positive constant, so that eventually, all the fleas will be carried as far to the right as we please.

We now claim that, conversely, if $\lambda < \frac{1}{n-1}$, then there is an initial configuration from which the fleas cannot be carried beyond a certain point M.

The positions of the fleas will be viewed as real numbers. Consider an arbitrary sequence of moves. Let S_k be the sum of all the numbers representing the positions of the fleas after the kth move and let w_k be the greatest of these numbers (i.e., the position of the rightmost flea). Note that $S_k \leq nw_k$. We are going to show that the sequence $\{w_k\}$ is bounded.

In the $(k+1)$th move a flea from a point A jumps over B, landing at C; let these points be represented by the numbers a, b, and c. Then $S_{k+1} = S_k + c - a$. By the given rules, $c - b = \lambda(b - a)$; equivalently, $\lambda(c - a) = (1 + \lambda)(c - b)$.

Thus, $S_{k+1} - S_k = c - a = \frac{1+\lambda}{\lambda}(c - b)$.

Suppose that $c > w_k$; the flea that has just jumped took the new rightmost position $w_{k+1} = c$. Since b was the position of some flea after the kth move, we have $b < w_k$ and

$$S_{k+1} - S_k = \frac{1+\lambda}{\lambda}(c - b) \geq \frac{1+\lambda}{\lambda}(w_{k+1} - w_k).$$

This estimate is also valid when $c \leq w_k$, in which case $w_{k+1} - w_k = 0$ and $S_{k+1} - S_k = c - a > 0$.

Consider the sequence of numbers

$$z_k = \frac{1+\lambda}{\lambda} \cdot w_k - S_k, \quad k = 0, 1, 2, \ldots,$$

so $z_{k+1} - z_k \leq 0$; the sequence is nonincreasing, and consequently $z_k \leq z_0$ for all k.

We have assumed that $\lambda < \frac{1}{n-1}$. Then $1 + \lambda > n\lambda$, and we can write

$$z_k = (n + \mu)w_k - S_k, \quad \text{where } \mu = \frac{1 + \lambda}{\lambda} - n > 0.$$

So, we get the inequality $z_k = \mu w_k + (nw_k - S_k) \geq \mu w_k$. It follows that $w_k \leq \frac{z_0}{\mu}$ for all k. Thus the position of the rightmost flea never exceeds a constant (depending on n, λ, and the initial configuration, but not on the strategy of moves).

In conclusion, the values of λ, asked about, are all real numbers not less than $\frac{1}{n-1}$.

【Score Situation】 This particular problem saw the following distribution of scores among contestants: 15 contestants scored 7 points, 1 contestant scored 6 points, 2 contestants scored 5 points, 6 contestants scored 4 points, 13 contestants scored 3 points, 29 contestants scored 2 points, 60 contestants scored 1 point, and 335 contestants scored 0 point. The average score of this problem is 0.655, indicating that it was extremely difficult.

Among the top five teams in the team scores, the scores of this problem are as follows: the China team scored 19 points (with a total team score of 218 points), the Russia team scored 25 points (with a total team score of 215 points), the United States team scored 27 points (with a total team score of 184 points), the South Korea team scored 15 points (with a total team score of 172 points), the Vietnam team scored 12 points (with a total team score of 169 points), and the Bulgaria team scored 5 points (with a total team score of 169 points).

The gold medal cutoff for this IMO was set at 30 points (with 39 contestants earning gold medals), the silver medal cutoff was 21 points (with 71 contestants earning silver medals), and the bronze medal cutoff was 11 points (with 119 contestants earning bronze medals).

In this IMO, a total of four contestants achieved a perfect score of 42 points.

Problem 4.8 (IMO 41-4, proposed by Hungary).
A magician has 100 cards numbered 1 to 100. He puts them into three boxes, a red one, a white one, and a blue one, so that each box contains at least one card. A member of the audience draws a card from each of two different boxes and announces the sum of the numbers on those cards.

Given this information, the magician locates the box from which no card has been drawn. How many ways are there to put the cards in the boxes so that this trick works?

Solution. Let the color of the number i be the color of the box which contains it. In the sequel, all numbers considered are assumed to be integers between 1 and 100.

Case 1: There is an i such that $i, i+1, i+2$ have three different colors, say **rwb**. Then, since $i + (i+3) = (i+1) + (i+2)$, the color of $i+3$ can be neither **w** (the color of $i+1$) nor **b** (the color of $i+2$). It follows that $i+3$ is **r**. We see that three neighbouring different colors determine the next one. Moreover, the pattern repeats: **rwb** is followed by **r**, then there come **w** and **b**, and so on. The argument works backwards as well: **rwb** is preceded by **b**, and so on.

So, it is enough to assign the colors of 1, 2, and 3, and this can be done in six different ways. All these arrangements are indeed good because the sums $\mathbf{r} + \mathbf{w}$, $\mathbf{w} + \mathbf{b}$, and $\mathbf{b} + \mathbf{r}$ give different remainders modulo 3.

Case 2: There are no three neighbouring numbers of different colors. Let 1 be red. Let i be the smallest non-red number, say, white. Let the smallest blue number be k. Since there is no **rwb**, we have $i+1 < k$.

Suppose that $k < 100$. Since $i + k = (i-1) + (k+1)$, the card $k+1$ should be red. However, in view of $i + (k+1) = (i+1) + k$, the card $i+1$ must be blue, which draws a contradiction to the fact that the smallest blue is k. This implies that k can only be 100.

Since $i + 99 = (i-1) + 100$, we see that 99 is white. We now show that 1 is red, 100 is blue, and all the others are white. If $t > 1$ were red, then in view of $t + 99 = (t-1) + 100$, the card $t-1$ should be blue, but the smallest blue is 100.

Hence, the coloring is $\mathbf{rww} \cdots \mathbf{wwb}$, and this is indeed good. If the sum is at most 100, then the missing box is blue; if the sum is 101, then it is white; if the sum is greater than 101, then it is red.

The number of such arrangements is again six. In conclusion, the answer is 12.

【Score Situation】This particular problem saw the following distribution of scores among contestants: 81 contestants scored 7 points, 49 contestants scored 6 points, 29 contestants scored 5 points, 28 contestants scored 4 points, 49 contestants scored 3 points, 62 contestants scored 2 points, 78 contestants scored 1 point, and 85 contestants scored 0 point. The average score of this problem is 3.182, indicating that it was relatively straightforward.

Among the top five teams in the team scores, the scores of this problem are as follows: the China team scored 39 points (with a total team score of 218 points), the Russia team scored 38 points (with a total team score of 215 points), the United States team scored 39 points (with a total team score of 184 points), the South Korea team scored 24 points (with a total team score of 172 points) , the Vietnam team scored 28 points (with a total team score of 169 points), and the Bulgaria team scored 25 points (with a total team score of 169 points).

The gold medal cutoff for this IMO was set at 30 points (with 39 contestants earning gold medals), the silver medal cutoff was 21 points (with 71 contestants earning silver medals), and the bronze medal cutoff was 11 points (with 119 contestants earning bronze medals).

In this IMO, a total of four contestants achieved a perfect score of 42 points.

Problem 4.9 (IMO 51-5, proposed by the Netherlands). In each of six boxes $B_1, B_2, B_3, B_4, B_5, B_6$ there is initially one coin. There are two types of operations allowed:

Type 1: Choose a nonempty box B_j with $1 \leq j \leq 5$. Remove one coin from B_j and add two coins to B_{j+1}.

Type 2: Choose a nonempty box B_k with $1 \leq k \leq 4$. Remove one coin from B_k and exchange the contents of (possibly empty) boxes B_{k+1} and B_{k+2}.

Determine whether there is a finite sequence of such operations that results in boxes B_1, B_2, B_3, B_4, B_5 being empty and box B_6 containing exactly $2010^{2010^{2010}}$ coins (note that $a^{b^c} = a^{(b^c)}$).

Solution. The answer is affirmative.

Let $A = 2010^{2010^{2010}}$. Denote by

$$(b_i, b_{i+1}, \ldots, b_{i+k}) \to (b_i', b_{i+1}', \ldots, b_{i+k}')$$

to mean that there is a finite sequence of operations on the boxes $B_i, B_{i+1}, \ldots, B_{i+k}$ initially containing $b_i, b_{i+1}, \ldots, b_{i+k}$ coins respectively that results in containing $b_i', b_{i+1}', \ldots, b_{i+k}'$ coins respectively. Then we shall show that

$$(1, 1, 1, 1, 1, 1) \to (0, 0, 0, 0, 0, A).$$

Lemma 1. *For every positive integer a, we have $(a, 0, 0) \to (0, 2^a, 0)$.*

Proof of Lemma 1. We prove by induction on $k \leq a$ that

$$(a, 0, 0) \to (a - k, 2^k, 0).$$

Since $(a, 0, 0) \to (a - 1, 2, 0) = (a - 1, 2^1, 0)$, the assertion is true for $k = 1$. Suppose that the assertion is true for some $k < a$. Then

$$(a-k, 2^k, 0) \to (a-k, 2^k-1, 2) \to \cdots \to (a-k, 0, 2^{k+1}) \to (a-k-1, 2^{k+1}, 0),$$

and thus

$$(a, 0, 0) \to (a - k, 2^k, 0) \to (a - k - 1, 2^{k+1}, 0).$$

The assertion is also true for $k + 1(\leq a)$. By the method of induction. Lemma 1 is proven.

Lemma 2. *For every positive integer a, we have $(a, 0, 0, 0) \to (0, P_a, 0, 0)$, where $P_n = 2^{2^{\cdot^{\cdot^{\cdot 2}}}}$ (with the n number of 2's) for a positive integer n.*

Proof of Lemma 2. We prove by induction on $k \leq a$ that

$$(a, 0, 0, 0) \to (a - k, P_k, 0, 0).$$

By the operation of type 1, we have

$$(a, 0, 0, 0) \to (a - 1, 2, 0, 0) = (a - 1, P_1, 0, 0),$$

and the assertion is true for $k = 1$. Suppose that the assertion is true for some $k < a$. Then

$$(a-k, P_k, 0, 0) \to (a-k, 0, 2^{P_k}, 0) = (a-k, 0, P_{k+1}, 0) \to (a-k-1, P_{k+1}, 0, 0),$$

and therefore

$$(a, 0, 0, 0) \to (a - k, P_k, 0, 0) \to (a - k - 1, P_{k+1}, 0, 0),$$

i.e., the assertion is also true for $k + 1(\leq a)$. By the method of induction, we have proven Lemma 2.

We have

$$(1, 1, 1, 1, 1, 1) \to (1, 1, 1, 1, 0, 3) \to (1, 1, 1, 0, 3, 0) \to (1, 1, 0, 3, 0, 0)$$
$$\to (1, 0, 3, 0, 0, 0) \to (0, 3, 0, 0, 0, 0) \to (0, 0, P_3, 0, 0, 0)$$
$$= (0, 0, 16, 0, 0, 0) \to (0, 0, 0, P_{16}, 0, 0)$$

and

$$A = 2010^{2010^{2010}} < (2^{11})^{2010^{2010}} = 2^{11 \cdot 2010^{2010}} < 2^{2010^{2011}} < 2^{(2^{11})^{2011}}$$
$$= 2^{2^{11 \cdot 2011}} < 2^{2^{15}} < P_{16},$$

and thus the number of coins in the box B_4 is greater than A. Therefore, by performing operations of type 2, we have

$$(0, 0, 0, P_{16}, 0, 0) \to (0, 0, 0, P_{16} - 1, 0, 0) \to (0, 0, 0, P_{16} - 2, 0, 0) \to \cdots$$
$$\to \left(0, 0, 0, \frac{A}{4}, 0, 0\right).$$

Then we get

$$\left(0, 0, 0, \frac{A}{4}, 0, 0\right) \to \cdots \to \left(0, 0, 0, 0, \frac{A}{2}, 0\right) \to \cdots \to (0, 0, 0, 0, 0, A).$$

【Score Situation】 This particular problem saw the following distribution of scores among contestants: 37 contestants scored 7 points, 11 contestants scored 6 points, 2 contestants scored 5 points, no contestant scored 4 points, 3 contestants scored 3 points, 26 contestants scored 2 points, 85 contestants scored 1 point, and 352 contestants scored 0 point. The average score of this problem is 0.932, indicating that it was extremely difficult.

Among the top five teams in the team scores, the scores of this problem are as follows: the China team scored 24 points (with a total team score of 197 points), the Russia team scored 20 points (with a total team score of 169 points), the United States team scored 24 points (with a total team score of 168 points), the South Korea team scored 0 point (with a total team score of 156 points), the Thailand team scored 5 points (with a total team score of 148 points), and the Kazakhstan team scored 4 points (with a total team score of 148 points).

The gold medal cutoff for this IMO was set at 27 points (with 47 contestants earning gold medals), the silver medal cutoff was 21 points (with 103 contestants earning silver medals), and the bronze medal cutoff was 15 points (with 115 contestants earning bronze medals).

In this IMO, only one contestant achieved a perfect score of 42 points, namely Zipei Nie from China.

Problem 4.10 (IMO 60-5, proposed by the United States). The Bank of Bath issues coins with an H on one side and a T on the other. Harry has n of these coins arranged in a line from left to right. He repeatedly performs the following operation: if there are exactly $k > 0$ coins showing H, then he turns over the kth coin from the left; otherwise, all coins show T and he stops. For example, when $n = 3$, the process starting with the configuration THT would be $THT \to HHT \to HTT \to TTT$, which stops after three operations.

(a) Show that, for each initial configuration, Harry stops after a finite number of operations.

(b) For each initial configuration C, let $L(C)$ be the number of operations before Harry stops. For example, $L(THT) = 3$ and $L(TTT) = 0$. Determine the average value of $L(C)$ over all 2^n possible initial configurations C.

Proof. Let $V_n = \{H, T\}^n$ be the set of all sequences of H and T with length n, i.e., the set of all statuses of the n coins. We apply induction to show that $L(C)$ is finite for any $C \in V_n$, and

$$\sum_{C \in V_n} L(C) = 2^{n-2} n(n+1).$$

For $n = 1$, we have $V_1 = \{H, T\}$, $L(H) = 1$, and $L(T) = 0$. For $n = 2$, we see that $V_2 = \{HH, HT, TH, TT\}$, $L(HH) = 2$, $L(HT) = 1$, $L(TH) = 3$, and $L(TT) = 0$. Thus, the assertion holds when $n = 1, 2$.

Suppose $n \geq 3$, and the assertion already holds for all smaller n. For two sequences A and B, let $A \cdot B$ denote the sequence obtained by concatenating the two sequences together (B is placed behind A). Let

$$X = \{C \cdot T \mid C \in V_{n-1}\}, \quad Y = \{H \cdot C \mid C \in V_{n-1}\},$$

$$Z = \{T \cdot C \cdot H \mid C \in V_{n-2}\}, \quad W = \{H \cdot C \cdot T \mid C \in V_{n-2}\}.$$

Apparently $V_n = X \cup Y \cup Z$ and $X \cap Y = W$.

For $C \cdot T \in X$, since the last term is T, operations on $C \cdot T$ are effectively the same as operations on C, so it becomes the sequence $TT \cdots T$ after $L(C)$ operations, i.e., $L(C \cdot T) = L(C)$.

For $H \cdot C \in Y$, suppose H appears k times in C. Then H appears $k + 1$ times in $H \cdot C$, and the $(k + 1)$th term in $H \cdot C$ is the same as the kth term in C. Hence, the operations on $H \cdot C$ are effectively first operating on C, to obtain $HTT \cdots T$ after $L(C)$ operations, and then operating on the first term to obtain a sequence of all T. Therefore, $L(H \cdot C) = L(C) + 1$.

For $T \cdot C \cdot H \in Z$, suppose H appears k times in C. Then H appears $k + 1$ times in $T \cdot C \cdot H$, and the $(k + 1)$th term in $T \cdot C \cdot H$ is the same as the kth term in C. Thus for $T \cdot C \cdot H$, we first operate on C, obtaining $TT \cdots TH$ after $L(C)$ operations, then change every T into H, from left to right, and finally change every H into T, from right to left. Hence, the process ends after $L(T \cdot C \cdot H) = L(C) + 2n - 1$ operations.

For $H \cdot C \cdot T \in W$, from the discussion in the previous cases we have

$$L(H \cdot C \cdot T) = L(H \cdot C) = L(C) + 1.$$

So far, we have shown that $L(C)$ is a finite number for every $C \in V_n$. Finally, applying the inductive hypothesis, we have

$$\sum_{C \in V_n} L(C) = \sum_{C \cdot T \in X} L(C \cdot T) + \sum_{H \cdot C \in Y} L(H \cdot C)$$

$$+ \sum_{T \cdot C \cdot H \in Z} L(T \cdot C \cdot H) - \sum_{H \cdot C \cdot T \in W} L(H \cdot C \cdot T)$$

$$= \sum_{C \in V_{n-1}} L(C) + \sum_{C \in V_{n-1}} (L(C) + 1)$$

$$+ \sum_{C \in V_{n-2}} (L(C) + 2n - 1) - \sum_{C \in V_{n-2}} (L(C) + 1)$$

$$= 2 \sum_{C \in V_{n-1}} L(C) + 2^{n-1} + (2n-2)2^{n-2}$$

$$= 2 \cdot 2^{n-3} n(n-1) + 2n \cdot 2^{n-2}$$

$$= 2^{n-2} n(n+1).$$

Therefore, the average of all $L(C)$ is $\frac{n(n+1)}{4}$.

【Score Situation】 This particular problem saw the following distribution of scores among contestants: 250 contestants scored 7 points, 3 contestants scored 6 points, 7 contestants scored 5 points, 5 contestants scored 4 points, 12 contestants scored 3 points, 168 contestants scored 2 points, 20 contestants scored 1 point, and 156 contestants scored 0 point. The average score of this problem is 3.567, indicating that it was relatively straightforward.

Among the top five teams in the team scores, the scores of this problem are as follows: the China team scored 27 points (with a total team score of 227 points), the United States team scored 26 points (with a total team score of 227 points), the South Korea team scored 26 points (with a total team score of 226 points), the North Korea team scored 17 points (with a total team score of 187 points), and the Thailand team scored 5 points (with a total team score of 185 points).

The gold medal cutoff for this IMO was set at 31 points (with 52 contestants earning gold medals), the silver medal cutoff was 24 points (with 94 contestants earning silver medals), and the bronze medal cutoff was 17 points (with 156 contestants earning bronze medals).

In this IMO, a total of six contestants achieved a perfect score of 42 points.

Problem 4.11 (IMO 63-1, proposed by Australia). The Bank of Oslo issues two types of coins: aluminium (denoted A) and bronze (denoted B). Marianne has n aluminium coins and n bronze coins, arranged in a row in some arbitrary initial order. A *chain* is any subsequence of consecutive coins of the same type. Given a fixed positive integer $k \le 2n$, Marianne repeatedly performs the following operation: she identifies the longest chain containing the kth coin from the left and moves all coins in that chain to the left end of the row. For example, when $n = 4$ and $k = 4$, the process starting from the ordering $AABBBABA$ would be

$$AABBBABA \to BBBAAABA \to AAABBBBA \to BBBBAAAA \to \cdots .$$

Find all pairs (n, k) with $1 \le k \le 2n$ such that for every initial ordering, at some moment during the process, the leftmost n coins will all be of the same type.

Solution. The desired are all pairs (n, k) that satisfy $n \le k \le \frac{3n+1}{2}$.

As defined in the problem, a chain is any subsequence of consecutive coins of the same type. We call it a "block," if it is a chain but not contained in a longer chain. For $M = A$ or B, let M^x represent a block of M coins of length x. We are interested in whether an initial ordering will turn into a 2-block sequence $A^n B^n$ or $B^n A^n$ after finitely many operations.

Claim 1. If $k < n$ or $k > \frac{3n+1}{2}$, then the sequence cannot always turn into $A^n B^n$ or $B^n A^n$.

Proof of the Claim 1. If $k < n$, then the sequence $A^{n-1} B^{n-1} A B$ does not change after an operation, and cannot turn into a 2-block sequence. If $k > \frac{3n+1}{2}$, then let $m = \lfloor \frac{n}{2} \rfloor$, $l = \lceil \frac{n}{2} \rceil$, and the initial ordering be $A^m B^m A^l B^l$. As $k > \frac{3n+1}{2} \geq m + l + l \geq m + m + l$, the kth coin always belongs to the rightmost block, and the process

$$A^m B^m A^l B^l \to B^l A^m B^m A^l \to A^l B^l A^m B^m \to B^m A^l B^l A^m$$
$$\to A^m B^m A^l B^l \to \cdots$$

is a 4-periodic cycle. It cannot turn into a 2-block sequence.

Claim 2. If $n \leq k \leq \frac{3n+1}{2}$, then any initial ordering will turn into a 2-block sequence.

Proof of the Claim 2. Note that after an operation, the number of blocks does not increase. Eventually, it will be stabilized, say at s blocks. It suffices to prove $s = 2$. Suppose $s > 2$. As $k \geq n$ and $s > 2$, the kth coin cannot belong to the leftmost block, and for the current operation, the block being moved is not the leftmost one.

If it is a middle one, then after the operation, the two blocks on its sides will merge into one block, reducing the number of blocks by 1, a contradiction. Therefore, when stabilized, at each operation the block being moved to the left is always the rightmost (sth) one.

Since $k \leq \frac{3n+1}{2}$, the length of the sth block is at least $2n - k + 1 \geq \frac{n+1}{2}$. Since s does not decrease after the operation, the first and the sth blocks are of different types, and hence s is even. Let $s = 2t$, $t \geq 2$, and suppose one of the stable orderings is $X_1 X_2 \cdots X_{2t}$ in which every X_i represents a block. The process is

$$X_1 X_2 \cdots X_{2t} \to X_{2t} X_1 \cdots X_{2t-1} \to \cdots X_2 X_2 \cdots X_1 \to X_1 X_2 \cdots X_{2t} \to \cdots .$$

Based on the previous argument, at any moment the rightmost block has length $\geq \frac{n+1}{2}$. So, each block has length $\geq \frac{n+1}{2}$, and the total length of the sequence is at least 4. Hence $\frac{n+1}{2} > 2n$, a contradiction. Now, the claim and the conclusion follow.

【Score Situation】This particular problem saw the following distribution of scores among contestants: 385 contestants scored 7 points, 51 contestants scored 6 points, 13 contestants scored 5 points, 10 contestants scored 4 points, 21 contestants scored 3 points, 19 contestants scored 2 points, 56 contestants scored 1 point, and 34 contestants scored 0 point. The average score of this problem is 5.540, indicating that it was simple.

Among the top five teams in the team scores, the scores of this problem are as follows: the China team scored 42 points (with a total team score of 252 points), the South Korea team scored 42 points (with a total team score of 208 points), the United States team scored 42 points (with a total team score of 207 points), the Vietnam team scored 40 points (with a total team score of 196 points), and the Romania team scored 42 points (with a total team score of 194 points).

The gold medal cutoff for this IMO was set at 34 points (with 44 contestants earning gold medals), the silver medal cutoff was 29 points (with 101 contestants earning silver medals), and the bronze medal cutoff was 23 points (with 140 contestants earning bronze medals).

In this IMO, a total of 10 contestants achieved a perfect score of 42 points.

4.2.3   Double-person operation problems

Problem 4.12 (IMO 31-5, proposed by Germany). Given an initial integer $n_0 > 1$ and two players A and B, choose integers $n_1, n_2, n_3, \ldots$, alternately according to the following rules:

Knowing n_{2k}, player A chooses any integer n_{2k+1} such that $n_{2k} \le n_{2k+1} \le n_{2k}^2$.

Knowing n_{2k+1}, player B chooses any integer n_{2k+2} such that $\frac{n_{2k+1}}{n_{2k+2}}$ is a prime raised to a positive integer power.

Player A wins the game by choosing the number 1990; player B wins by choosing the number 1. For which n_0 does:

(a) Player A has a winning strategy?
(b) Player B has a winning strategy?
(c) Neither player has a winning strategy?

Solution. If $n_0 = 2$, then A can choose only from 2, 3, and 4, and B can choose 1 and wins.

If $n_0 = 3$, then A can choose only from 3 to 9, all of the form p^k or $2p^k$ with p prime, and B can choose 1 or 2. Hence B has a winning strategy.

If $n_0 = 4$, then A can choose only from 4 to 16, all of the form p^k, $2p^k$, or $3p^k$, and B can choose 1, 2, or 3. Hence B has a winning strategy.

If $n_0 = 5$, then A can choose only from 5 to 25, all of the form p^k, $2p^k$, $3p^k$, or $4p^k$, and B can choose 1, 2, 3, or 4. Hence B has a winning strategy.

If $45 \le n_0 \le 1990$, then A can choose 1990 immediately and wins.

If $21 \le n_0 \le 44$, then A chooses $420 = 2^2 \times 3 \times 5 \times 7$, which forces B to choose a number between 45 and 1990. Hence A has a winning strategy.

If $13 \le n_0 \le 20$, then A chooses $n_1 = 2^3 \times 3 \times 7 = 168$, which forces B to choose a number between 21 and 1990. Hence A has a winning strategy.

If $11 \le n_0 \le 12$, then A chooses $n_1 = 3 \times 5 \times 7 = 105$, which forces B to choose a number between 15 and 1990. Hence A has a winning strategy.

If $8 \le n_0 \le 10$, then A chooses $n_1 = 2^2 \times 3 \times 5 = 60$, which forces B to choose a number between 12 and 1990. Hence A has a winning strategy.

If $n_0 > 1990$, then A chooses $n_1 = 2^{r+1} \times 3^2$ such that

$$2^r \times 3^2 < n_0 \le 2^{r+1} \times 3^2 < n_0^2.$$

The number n_2 which B chooses must satisfy $8 \le n_2 < n_0$. If $n_2 > 1990$, substitute n_2 for n_0 and proceed as above. After a finite number of steps, we arrive at $8 \le n_{2k} \le 1990$, and hence A has a winning strategy.

Finally, we are left with the case $n_0 = 6$ or 7. If A chooses $n_1 = 30$, then B can only choose 6 (for otherwise B chooses 10 or 15, and we know A has a winning strategy). Thus, A can choose 30 again, so A has a non-losing strategy.

On the other hand, among all numbers ≤ 49, A can only choose 30 or 42 in order to make sure that $n_2 \ge 6$ (for all other choices of $n_1 \le 49$, B can choose $n_2 < 6$ and wins), now B can choose $n_2 = 6$, and this procedure repeats. Thus, B also has a non-losing strategy. If $n_0 = 6$ or 7, then neither has a winning strategy.

In conclusion, if $n_0 \ge 8$, then A has a winning strategy; if $n_0 \le 5$, then B has a winning strategy; If $n_0 = 6$ or 7, then neither has a winning strategy.

【Score Situation】 This particular problem saw the following distribution of scores among contestants: 100 contestants scored 7 points, 26 contestants scored 6 points, 26 contestants scored 5 points, 22 contestants scored 4 points, 31 contestant scored 3 points, 43 contestants scored 2 points, 39 contestants scored 1 point, and 21 contestants scored 0 point. The average score of this problem is 4.195, indicating that it was simple.

Among the top five teams in the team scores, the scores of this problem are as follows: the China team scored 40 points (with a total team score of 230 points), the Soviet Union team scored 40 points (with a total team score of 193 points), the United States team scored 39 points (with a total team score of 174 points), the Romania team scored 35 points (with a total team score of 171 points), and the France team scored 24 points (with a total team score of 168 points).

The gold medal cutoff for this IMO was set at 34 points (with 23 contestants earning gold medals), the silver medal cutoff was 23 points (with 56 contestants earning silver medals), and the bronze medal cutoff was 16 points (with 76 contestants earning bronze medals).

In this IMO, a total of four contestants achieved a perfect score of 42 points.

Problem 4.13 (IMO 53-3, proposed by Canada). A *liar's guessing game* is a game played between two players A and B. The rule of the game depends on two positive integers k and n which are known to both players.

At the start of the game, A chooses integers x and N with $1 \leq x \leq N$. Player A keeps x secret, and truthfully tells N to player B. Player B now tries to obtain information about x by asking player A questions as follows: each question consists of B specifying an arbitrary set S of positive integers (possibly one specified in some previous question), and asking A whether x belongs to S. Player B may ask as many such questions as he wishes. After each question, player A must immediately answer it with *yes* or *no* but is allowed to lie as many times as she wants; the only restriction is that, among any $k+1$ consecutive answers, at least one answer must be truthful.

After B has asked as many questions as he wants, he must specify a set X of at most n positive integers. If x belongs to X, then B wins; otherwise, he loses. Prove that:

(1) If $n \geq 2^k$, then B can guarantee a win.
(2) For all sufficiently large k, there exists an integer $n \geq 1.99^k$ such that B cannot guarantee a win.

Proof. We rephrase the rule of the game as follows: Given two positive integers k and n, player A tells a finite set $T = \{1, 2, \ldots, N\}$ to player B and keeps one element x secret. Player B chooses a finite subset S of T and asks player A whether x belongs to S. Player A answers it with yes or no and is allowed to lie at most consecutive k times. After asking a finite number of questions, if player B can specify a subset of T containing x with at most n elements, then player B wins.

(1) If $N > 2^k$, then we show that player B can always determine some element $y \in T$ such that $y \neq x$. Thus, we can restrict $N \leq 2^k$. Consequently, if $n \geq 2^k$, then player B wins.

Denote $T = \{1, 2, \ldots, N\}$. Player B asks player A the same question k times: Is $x = 2^k + 1$? If the answers are all no, then $x \neq 2^k + 1 = y$.

Once an answer is yes, then the next first question of player B is: Is $x \in \{t \in \mathbf{Z} \mid 1 \leq t \leq 2^{k-1}\}$?

If A answers yes, then we must find a $y \in \{t \in \mathbf{Z} \mid 2^{k-1} + 1 \leq t \leq 2^k\}$ such that $y \neq x$. If A answers no, then we must find a $y \in \{t \in \mathbf{Z} \mid 1 \leq t \leq 2^{k-1}\}$ such that $y \neq x$.

In this way, by each answer of player A, we can reduce half the size of the set containing y. When A answers k questions, we can conclude that there is a unique a with $1 \leq a \leq 2^k$. If $a = x$, then A lies consecutive $k + 1$ times. So $y = a \neq x$.

(2) We prove that for any $1 < \lambda < 2$, if $n = [(2 - \lambda)\lambda^{k+1}] - 1$, then player B cannot guarantee a win. Specially, take λ with $1.99 < \lambda < 2$, and for sufficiently large integer k, we have

$$n = [(2 - \lambda)\lambda^{k+1}] - 1 > 1.99^k,$$

which is the required conclusion.

Player A chooses $T = \{1, 2, \ldots, n+1\}$ and chooses any $x \in T$. Denote the maximum number of consecutive lies by m_i when $x = i$. And define $\phi = \sum_{i=1}^{n+1} \lambda^{m_i}$. The strategy of A is to choose to lie or not such that he takes ϕ the smaller value.

We will show that $\phi < \lambda^{k+1}$ at any time with the strategy stated above. Thus $m_i \leq k$ for each $i \in T$. So, B cannot determine whether $i = x$ or not. Thus, player B cannot guarantee a win.

In the following we show that $\phi < \lambda^{k+1}$. At the beginning, $m_i = 0$, so $\phi = n + 1 < \lambda^{k+1}$. Suppose that after several answers, $\phi < \lambda^{k+1}$, and B asks: Is $x \in S$? The answer "yes" or "no" corresponding to two numbers of ϕ, respectively, is as follows: $\phi_1 = \sum_{i \in S} 1 + \sum_{i \notin S} \lambda^{m_i+1}$ and $\phi_2 = \sum_{i \notin S} 1 + \sum_{i \in S} \lambda^{m_i+1}$. By definition,

$$\phi = \min(\phi_1, \phi_2) \leq \frac{1}{2}(\phi_1 + \phi_2) = \frac{1}{2}(\lambda\phi + n + 1)$$

$$< \frac{1}{2}(\lambda^{k+2} + (2 - \lambda)\lambda^{k+1}) = \lambda^{k+1}.$$

【Score Situation】This particular problem saw the following distribution of scores among contestants: 8 contestants scored 7 points, no contestant scored 6 points, 6 contestants scored 5 points, 7 contestants scored 4 points, 31 contestants scored 3 points, 4 contestants scored 2 points, 11 contestants scored 1 point, and 480 contestants scored 0 point. The average score of this problem is 0.413, indicating that it was extremely difficult.

Among the top five teams in the team scores, the scores of this problem are as follows: the South Korea team scored 21 points (with a total team score of 209 points), the China team scored 14 points (with a total team score of 195 points), the United States team scored 33 points (with a total team score of 194 points), the Russia team scored 21 points (with a total team score of 177 points), the Canada team scored 9 points (with a total team score of 159 points), and the Thailand team scored 4 points (with a total team score of 159 points).

The gold medal cutoff for this IMO was set at 28 points (with 51 contestants earning gold medals), the silver medal cutoff was 21 points (with 88 contestants earning silver medals), and the bronze medal cutoff was 14 points (with 137 contestants earning bronze medals).

In this IMO, only one contestant achieved a perfect score of 42 points, namely Jeck Lim from Singapore.

Problem 4.14 (IMO 58-3, proposed by Austria). A hunter and an invisible rabbit play a game in the Euclidean plane. The rabbit's starting point, A_0, and the hunter's starting point, B_0, are the same. After $n - 1$ rounds of the game, the rabbit is at point A_{n-1} and the hunter is at point B_{n-1}. In the nth round of the game, three things occur in order.

(i) The rabbit moves invisibly to a point A_n such that the distance between A_{n-1} and A_n is exactly 1.

(ii) A tracking device reports a point P_n to the hunter. The only guarantee provided by the tracking device to the hunter is that the distance between P_n and A_n is at most 1.

(iii) The hunter moves visibly to a point B_n such that the distance between B_{n-1} and B_n is exactly 1.

Is it always possible, no matter how the rabbit moves, and no matter what points are reported by the tracking device, for the hunter to choose her moves so that after 10^9 rounds she can ensure that the distance between her and the rabbit is at most 100?

Solution. The hunter cannot guarantee that after 10^9 rounds the distance between her and the rabbit is at most 100.

First, if the solution to the problem is "the hunter has a strategy that meets the requirements of the problem," then this strategy should be applicable to any sequence of P_n (and any sequence of A_n that satisfies Condition (ii)). We will explain that if the hunter has a "bad luck," then she cannot guarantee this.

Let $d_n = A_n B_n$. Suppose there exists $n \leq 10^9$ such that $d_n \geq 100$. Then after this, the rabbit just needs to jump along the line between it and the hunter, in the opposite direction to the latter.

Now suppose $d_n < 100$. We prove that when the rabbit moves in an appropriate way and the feedback points of the positioning device are "beneficial" for the rabbit, no matter what kind of movement the hunter takes, after 200 rounds, there is no guarantee that $d_{n+200}^2 < d_n^2 + \frac{1}{2}$, or in other words, it is always possible that $d_{n+200}^2 \geq d_n^2 + \frac{1}{2}$. Then for $d_0 = 0$ at the beginning, after $n_0 = 2 \cdot 10^4 \cdot 200 = 4 \cdot 10^6 < 10^9$ rounds, the rabbit was able to successfully increase the distance between it and the hunter to at least 100.

Assume that after the nth round, the rabbit is at A_n and the hunter is at B_n. We can even let the rabbit show its position to the hunter (this makes the previous feedback information from the positioning device negligible). Let l be the straight line $A_n B_n$, and Y_1 and Y_2 be two points on both sides of l, whose distances to A_n be both 200 (their distances to B_n be greater than 200) and that to l be both 1. As shown in Figure 4.5.

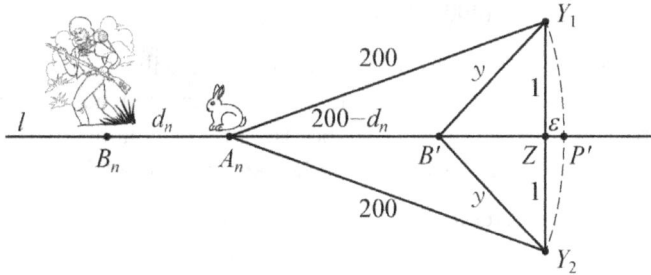

Figure 4.5

The rabbit's strategy is to choose one of Y_1 and Y_2 as the target and jump towards it along the straight line linking it and A_n. At the same time, suppose the points returned by the positioning device all fall on line l. In this way, the hunter will not know whether the rabbit jumped to Y_1 or Y_2.

In other words, we assume that the points of the next 200 positioning device feedback are exactly 200 points that advance a distance of 1 each time in the direction of the ray on l starting from A_n, that is,

$$A_n P_{n+1} = P_{n+1} P_{n+2} = \cdots = P_{n+199} P_{n+200} = 1.$$

Denote $P' = P_{n+200}$. We may let the hunter know the next 200 feedback points $P_{n+1}, P_{n+2}, \ldots, P_{n+200}$ in the $n+1$ round beforehand, which will only increase the amount of information the hunter has.

How does the hunter decide after seeing these 200 feedback points? If the hunter moves 200 steps from B_n along $B_n P_{n+200}$, then he will reach

B', and $B_n B' = 200$. Please note that the hunter has no better strategy! Because, after the hunter moves 200 steps according to any strategy, his position will always be the left side of B'; if the position is above l, then her distance to Y_2 will become larger. Otherwise, her distance to Y_1 will become larger. In short, no matter how the hunter chooses her movement method, she cannot guarantee that the distance from the rabbit to her will be less than $y = B'Y_1 = B'Y_2$.

To estimate the value of y^2, let Z be the midpoint of segment $Y_1 Y_2$ and $\varepsilon = ZP'$ (note that $B'P' = d_n$). We have

$$\varepsilon = 200 - A_n Z = 200 - \sqrt{200^2 - 1} = \frac{1}{200 + \sqrt{200^2 + 1}} > \frac{1}{400}.$$

Particularly, $\varepsilon^2 + 1 = 400\varepsilon$. Then, assuming $d_n < 100$, we have

$$y^2 = 1 + B'Z^2 = 1 + (d_n - \varepsilon)^2 = d_n^2 - 2\varepsilon d_n + \varepsilon^2 + 1$$

$$= d_n^2 + \varepsilon(400 - 2d_n) > d_n^2 + \frac{1}{2}.$$

Therefore, as long as the positioning device feedback gives such a list of points $P_{n+1}, P_{n+2}, \ldots, P_{n+200}$, no matter what the hunter does, it is possible to make $d_{n+200}^2 > d_n^2 + \frac{1}{2}$. So, the hunter cannot guarantee that after 10^9 rounds the distance between her and the rabbit is at most 100.

【Score Situation】 This particular problem saw the following distribution of scores among contestants: 2 contestants scored 7 points, no contestant scored 6 points, 1 contestant scored 5 points, 1 contestant scored 4 points, no contestant scored 3 points, no contestant scored 2 points, 3 contestants scored 1 point, and 608 contestants scored 0 point. The average score of this problem is 0.042, indicating that it was extremely difficult.

Among the top five teams in the team scores, the scores of this problem are as follows: the South Korea team scored 1 points (with a total team score of 170 points), the China team scored 0 point (with a total team score of 159 points), the Vietnam team scored 0 point (with a total team score of 155 points), the United States team scored 0 point (with a total team score of 148 points), and the Iran team scored 0 point (with a total team score of 142 points).

The gold medal cutoff for this IMO was set at 25 points (with 48 contestants earning gold medals), the silver medal cutoff was 19 points (with 90 contestants earning silver medals), and the bronze medal cutoff was 16 points (with 153 contestants earning bronze medals).

In this IMO, no contestant achieved a perfect score of 42 points.

Problem 4.15 (IMO 59-4, proposed by Armenia). A *site* is any point (x, y) in a plane such that x and y are both positive integers less than or equal to 20.

Initially, each of the 400 sites is unoccupied. Amy and Ben take turns placing stones with Amy going first. On her turn, Amy places a new red stone on an unoccupied site such that the distance between any two sites occupied by red stones is not equal to $\sqrt{5}$. On his turn, Ben places a new blue stone on any unoccupied site. (A site occupied by a blue stone is allowed to be at any distance from any other occupied site.) They stop as soon as a player cannot place a stone.

Find the greatest K such that Amy can ensure that she places at least K red stones, no matter how Ben places his blue stones.

Solution. $K = 100$.

We first claim that Amy has a strategy to place at least 100 red stones. Classify all the sites into two groups, odd or even. If $2 \nmid (x + y)$, then we call the site (x, y) odd; otherwise, we call it even. Since the distance between any two odd sites is not $\sqrt{5}$, and there are 200 odd sites, Amy can place 100 red stones in her first 100 moves on unoccupied odd sites. This proves our Claim 1.

We next claim that Ben has a strategy to prevent Amy from placing more than 100 red stones. Consider a 4×4 array of sites. Divide the 16 sites into four groups, as shown in Figure 4.6; the sites labeled the same letter belong to the same group.

$$
\begin{array}{cccc}
A & B & C & D \\
C & D & A & B \\
B & A & D & C \\
D & C & B & A
\end{array}
$$

Figure 4.6

If we connect two sites with distance $\sqrt{5}$ in the same group, then each group forms a parallelogram. Ben first divides the 400 sites into the 25 number of 4×4 arrays, each array is further divided to four groups, and each group is connected to from a parallelogram as above. Now Ben has 100 parallelograms.

If Amy places a red stone on the vertex P of some parallelogram, then Ben immediately places a blue stone on the opposite vertex of that parallelogram, preventing Amy from placing red stones on the two adjacent vertices of P. Hence Amy can place at most one red stone on each parallelogram. This proves our Claim 2.

In conclusion, the asked $K = 100$.

【Score Situation】 This particular problem saw the following distribution of scores among contestants: 271 contestants scored 7 points, 5 contestants scored 6 points, 15 contestants scored 5 points, 18 contestants scored 4 points, 18 contestants scored 3 points, 106 contestants scored 2 points, 13 contestants scored 1 point, and 148 contestants scored 0 point. The average score of this problem is 3.961, indicating that it was relatively straightforward.

Among the top five teams in the team scores, the scores of this problem are as follows: the United States team scored 42 points (with a total team score of 212 points), the Russia team scored 42 points (with a total team score of 201 points), the China team scored 42 points (with a total team score of 199 points), the Ukraine team scored 42 points (with a total team score of 186 points), and the Thailand team scored 42 points (with a total team score of 183 points).

The gold medal cutoff for this IMO was set at 31 points (with 48 contestants earning gold medals), the silver medal cutoff was 25 points (with 98 contestants earning silver medals), and the bronze medal cutoff was 16 points (with 143 contestants earning bronze medals).

In this IMO, only two contestants achieved a perfect score of 42 points, namely Agnijo Banerjee from the United Kingdom and James Lin from the United States.

Problem 4.16 (IMO 62-5, proposed by Spain). Two squirrels, Bushy and Jumpy, have collected 2021 walnuts for the winter. Jumpy numbers the walnuts from 1 through 2021 and digs 2021 little holes in a circular pattern in the ground around their favourite tree. The next morning Jumpy notices that Bushy had placed one walnut into each hole but had paid no attention to the numbering. Unhappy, Jumpy decides to reorder the walnuts by performing a sequence of 2021 moves. In the kth move, Jumpy swaps the positions of the two walnuts adjacent to walnut k.

Prove that there exists a value of k such that on the kth move, Jumpy swaps some walnuts a and b such that $a < k < b$.

Proof 1. Assume that the statement is untrue. In the kth move, if Jumpy swaps a and b with $a, b < k$, then we say k is a "big" walnut; if $a, b > k$, then we say k is a "small" walnut. Before the kth move, we call every walnut with number $a < k$ "operated" and every walnut with number $b > k$ "unoperated."

At every moment, the operated walnuts are divided by the unoperated walnuts into several intervals (each interval consists of consecutive operated walnuts and the neighbors on the two ends are unoperated ones). We use induction to prove that, after k moves, each interval contains an odd number of walnuts. For $k = 1$, the assertion is obvious. In the kth move, there are two possibilities:

(i) If k is a small walnut, then its neighbors are both unoperated before and after the move, and k itself is an interval of length 1.
(ii) If k is a big walnut, then its neighbors are both operated, say they belong to intervals of lengths p and q. After the move, they merge into one interval of length $p + q + 1$. By the inductive hypothesis, p and q are both odd, and the new interval's length $p + q + 1$ is also odd.

Now the induction is complete, and the assertion is verified. Notice that after 2020 moves, the 2020 operated walnuts form a single interval of length 2020 which is an even number. This contradiction indicates that during the process, Jumpy must have swapped walnuts a and b adjacent to k such that $a < k < b$.

Proof 2. As in the first proof, we assume that the statement is untrue and define the big and small walnuts in the same way. In the kth move, the walnut with number k is called *being operated*, and the two adjacent walnuts are called *being swapped*.

Claim 3. If k is a small walnut, then in the kth move, Jumpy swaps two big walnuts.

Claim 4. If k is a big walnut, then in the kth move, Jumpy swaps two small walnuts.

Proof of Claim 3. Assume that in the kth move, a small walnut a is swapped. Clearly $a > k$ since k is small. Before the ath move, a has been swapped (for example in the kth move).

Assume the last time a is swapped before the ath move is in the k'th move. Then $a > k' \geq k$ and k' is also a small walnut. Before and after the k'th move, a is adjacent to k'. After the k'th move, let walnut k' be in the hole P. Walnut a is not swapped after the k'th move and before the ath move. However, in the ath move, the walnut in hole P is numbered bigger than a since a is small.

Thus, in some move after the k'th and before the ath one, the walnut in the hole P is changed from an operated one to an unoperated one. Let it be the lth move with $k' < l < a$. In the lth move, the two swapped walnuts are an operated one (which is numbered smaller than l) and an unoperated one (which is numbered larger than l), contrary to the assumption.

Proof of Claim 4. Claim 4 follows from Claim 3 if we reverse the time and treat walnut i as walnut $2022 - i$. The moves are proceeded reversely starting from the final configuration and finishing at the initial configuration. However, the big walnuts are the original small walnuts, and the small walnuts are the original big walnuts. And Claim 3 in the reversal moves implies Claim 4 in the original moves.

From Claim 3 and Claim 4, we see that two swapped walnuts at each move are both big or both small. In particular, the size of walnuts in each hole remains unchanged during the moves. Thus, we may define a hole to be a big hole if the walnut in it is big, or a small hole otherwise. We shall prove that the adjacent holes of a big hole are small, and the adjacent holes of a small hole are big. Then it follows that the big holes and small holes are arranged on the circle alternatively. However, 2021 is odd, so this is impossible.

To prove that the two adjacent holes of a small hole are big, consider a small hole P which initially contains a small walnut a. Walnut a cannot be swapped before ath move; otherwise say a is swapped in the kth move where $k < a$, then k is small and a is big by Claim 3, which is absurd.

So, a remains in P until the ath move when it is operated. Then the two adjacent holes are big since a is small. The fact that the two adjacent holes of a big hole are small again follows from the reversal argument as in the proof of Claim 4. This completes the proof.

【Score Situation】 This particular problem saw the following distribution of scores among contestants: 175 contestants scored 7 points, 4 contestants scored 6 points, 5 contestants scored 5 points, 2 contestants scored 4 points, 4 contestants scored 3 points, 13 contestants scored 2 points, 12 contestants scored 1 point, and 404 contestants scored 0 point. The average score of this problem is 2.152, indicating that it had a certain level of difficulty.

Among the top five teams in the team scores, the scores of this problem are as follows: the China team scored 42 points (with a total team score of 208 points), the Russia team scored 35 points (with a total team score of 183 points), the South Korea team scored 35 points (with a total team score of 172 points), the United States team scored 35 points (with a total team score of 165 points), and the Canada team scored 42 points (with a total team score of 151 points).

The gold medal cutoff for this IMO was set at 24 points (with 52 contestants earning gold medals), the silver medal cutoff was 19 points (with 103 contestants earning silver medals), and the bronze medal cutoff was 12 points (with 148 contestants earning bronze medals).

In this IMO, only one contestant achieved a perfect score of 42 points, namely Yichuan Wang from China.

4.3 Summary

In the first 64 IMOs, there were a total of 16 operation and logical reasoning problems. These problems can be broadly categorized into three types, as depicted in Figure 4.7. The score details for these problems are presented in Table 4.3. Due to the smaller number of participating teams and missing contestant score information in early IMOs, there are several blanks in Table 4.3.

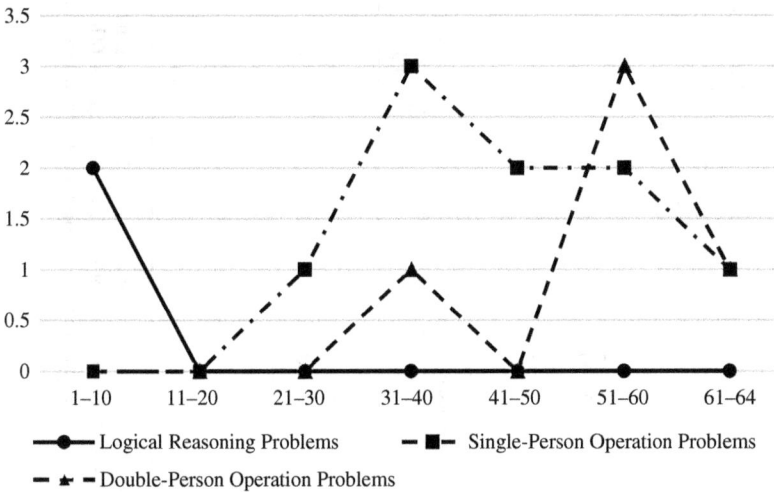

Figure 4.7 Numbers of Operation and Logical Reasoning Problems in the First 64 IMOs

Problems 4.1 and 4.2 focus on "logical reasoning problems;" among these two problems, the one with the lowest average score is Problem 4.2 (IMO 16-1), proposed by the United States. Problems 4.3–4.11 deal with "single-person operation problems;" among these nine problems, the one with the lowest average score is Problem 4.7 (IMO 41-3), proposed by Belarus. Problems 4.12–4.16 are about "double-person operation problems;" among these five problems, the one with the lowest average score is Problem 4.14 (IMO 58-3), proposed by Austria.

These 16 problems were proposed by 12 countries. The United States, Hungary, Finland, and the Netherlands each contributed two problems.

From Table 4.3, it can be observed that in the first 64 IMOs, there were five operation and logical reasoning problems with an average score of 0–1 point; two problems with an average score of 1–2 points; one problem with

Table 4.3   Score Details of Operation and Logical Reasoning Problems in the First 64 IMOs

Problem	4.1	4.2	4.3	4.4	4.5	4.6	4.7	4.8
Full points	8	5	7	7	7	7	7	7
Average score	6.063	4.514	0.871	1.131	1.816	3.175	0.655	3.182
Top five mean		5.000	2.500	4.100	5.133	5.133	3.267	5.600
6th–15th mean		4.636	1.217	1.909	2.955	4.850	1.426	4.444
16th–25th mean			0.293	1.370	2.481	3.727	0.867	3.983
Problem number in the IMO	5-6	16-1	27-3	34-3	34-6	37-1	41-3	41-4
Proposing country	Hungary	The United States	The German Democratic Republic	Finland	The Netherlands	Finland	Belarus	Hungary

Problem	4.9	4.10	4.11	4.12	4.13	4.14	4.15	4.16
Full points	7	7	7	7	7	7	7	7
Average score	0.932	3.567	5.540	4.195	0.413	0.042	3.961	2.152
Top five mean	2.433	3.367	6.933	5.933	3.267	0.033	7.000	6.300
6th–15th mean	1.574	6.652	6.933	5.939	0.833	0.258	6.533	5.150
16th–25th mean	1.350	5.593	6.867	4.259	0.750	0.000	6.183	4.583
Problem number in the IMO	51-5	60-5	63-1	31-5	53-3	58-3	59-4	62-5
Proposing country	The Netherlands	The United States	Australia	Germany	Canada	Austria	Armenia	Spain

Note. Top five mean = Total score of the top five teams ÷ Total number of contestants from the top five teams,

6th–15th mean = Total score of the 6th–15th teams ÷ Total number of contestants from the 6th–15th teams,

16th–25th mean = Total score of the 16th–25th teams ÷ Total number of contestants from the 16th–25th teams.

an average score of 2–3 points; four problems with an average score of 3–4 points; four problems with an average score above 4 points.

In the 24th–64th IMOs, there were a total of 14 operation and logical reasoning problems. Among these, five had an average score of 0–1 point; two had an average score of 1–2 points; one had an average score of 2–3 points; four had an average score of 3–4 points; two had an average score above 4 points. Further analysis of the problem numbers of these 14 operation and logical reasoning problems, as shown in Table 4.4, reveals that the numbers of operation problems of different difficulty levels are relatively close.

Table 4.4 Numbers of Operation and Logical Reasoning Problems in the 24th–64th IMOs

Operation and Logical Reasoning Problem	Problem Number			Number of Problems in the First 64 IMOs
	1, 4	2, 5	3, 6	
Logical reasoning problems	0	0	0	2
Single-person operation problems	3	2	4	9
Double-person operation problems	1	2	2	5
Total	4	4	6	16

From Table 4.3, it is observable that in the operation and logical reasoning problems, the average score of the top five teams generally exceeds the average score of the problem by 2.5 points, the average score of the 6th–15th teams typically surpasses the average score by 1 point, and the average score of the 16th–25th teams usually exceeds the average score by 0.5 points.

It can also be observed that there is an operation problem where the average score is higher than the average score of the 16th–25th teams, such as Problem 4.3 (IMO 27-3). This phenomenon is due to the smaller number of participating teams in early IMOs. It was not until the 30th IMO in 1989 that the number of participating teams exceeded 50. Therefore, it is common to see situations where the average socre is close to or even higher than the average score of the 16th–25th teams during this period.

Chapter 5

Combinatorial Geometry Problems

The term "combinatorial geometry" originated from a monograph published in 1955 by H. Hadwiger and others titled "Der Kombinatorischen Geometrie in der Ebene" (Combinatorial Geometry in the Plane). It is internationally recognized by the mathematical community that this publication marked the emergence of combinatorial geometry as a new discipline. However, mathematicians like Euler and Kepler explored combinatorial geometry long before this work.

The development of computer science has also injected new vitality into the field of combinatorial geometry. In 1998, Thomas Hales combined a 250-page mathematical proof with extensive computer calculations to finally prove the "Kepler Conjecture" regarding sphere packing. Subsequently, in 2016, the mathematician Maryna Viazovska solved two high-dimensional cases of sphere-packing problems, for which she was awarded a Fields Medal in 2022.

Unlike other combinatorial problems, combinatorial geometry focuses on more specific geometric figures, studying their positions, shapes, and other properties. Combinatorial geometry problems also appear frequently in the IMO.

In the first 64 IMOs, there had been a total of 27 combinatorial geometry problems, accounting for approximately 31.0% of all combinatorics problems. These problems can be primarily categorized into four types: (1) point and line problems, totaling 17 problems; (2) shape problems, totaling three problems; (3) covering, embedding, partitioning, and patching problems, totaling two problems; (4) lattice point and grid problems,

totaling five problems. The statistical distribution of these four types of
problems in the previous IMOs is presented in Table 5.1.

**Table 5.1 Numbers of Combinatorial Geometry Problems in the First
64 IMOs**

Content	Session							Total
	1–10	11–20	21–30	31–40	41–50	51–60	61–64	
Point and line problems	2	3	4	2	0	5	1	17
Shape problems	0	1	0	1	1	0	0	3
Covering, embedding, partitioning, and patching problems	0	1	0	0	1	0	0	2
Lattice point and grid problems	0	0	1	1	0	2	1	5
Combinatorics problems	7	11	16	16	11	19	7	87
Percentage of combinatorial geometry problems among combinatorics problems	28.6%	45.5%	31.3%	25.0%	18.2%	36.8%	28.6%	31.0%

The problem of combinatorial geometry has frequently appeared in the
IMO from early to recent years, with the most common being the problem
concerning configurations of points and lines. There are also many properties and theorems about the set of points and lines in a plane.

It is important to note that for each problem, the solutions are followed by information on the scores, including the number of contestants in
each score range, the average score, and the scores of the top five teams.
However, early IMOs often lacked information on contestant scores, so the
number of contestants in each score range only represents the counted number of contestants, and some problems lack scores of the top five teams.

5.1 Common Theorems, Formulas and Methods

5.1.1 *Preliminaries*

(1) *Partition, composition, and other aspects of graphics*

Let A be a planar point set. A point X in the plane is called an accumulation point of A if an arbitrarily small neighborhood of X contains

some point in A. Note that accumulation points may not belong to A; for example when A is the open interval (a, b), any point in the closed interval $[a, b]$ is an accumulation point of A.

The set A together with all its accumulation points is call the closure of A, denoted by $\text{cl}(A)$. If $\text{cl}(A) = A$, then A is called a closed set. Intuitively, a geometric figure that includes its boundary is a closed set.

If a point set is divided into pairwise disjoint point sets, then it is called a "point set partition." If a geometric region is divided into closed sets with no common interior points (which may only have common points on the boundary), then it can be called a "geometric partition."

If we can partition two point sets A and B into several parts respectively, say $A_1, A_2, \ldots, A_n$ and $B_1, B_2, \ldots, B_n$, such that A_i and B_i $(i = 1, 2, \ldots, n)$ are congruent (resp., similar), then A and B are called composedly congruent (resp., composedly similar). If the partitions are geometric partitions, then A and B are called geometrically congruent (resp., similar).

(2) Geometric properties of lattice points

Theorem 5.1 (Minkowski Theorem). *If a planar convex body is symmetric about the origin with area larger than 4, then there exists a lattice point other than the origin in the interior.*

Theorem 5.2 (Pick's Theorem). *Assume that a lattice polygon (not necessarily convex) has m lattice points in the interior and n lattice points on the boundary. Then its area is equal to $m + \frac{n}{2} - 1$.*

(3) Connectivity and convex sets

A point set A is called connected if for any two points in A, there is a curve connecting these two points lying in A.

Projecting the shapes onto a line is a very useful trick.

Theorem 5.3. *Given a finite number of connected regions (not necessarily convex) in a plane, where every two have a common point, there exists a straight line passing through all of these regions.*

For two points U and V in a point set, if the segment UV lies entirely in the point set, then we say that $U(V)$ is visible to $V(U)$. If any two points are visible to each other, then we say the set A is convex. The boundary of a convex set is denoted by ∂A. Note that points in ∂A may not belong to A. If ∂A is contained in A, then A is closed. A bounded closed convex set is called a convex body.

We use $d(U, V)$ to denote the distance between U and V. The diameter of a point set A is defined to be $\sup d(U, V)$, where $U, V \in A$. The distance between two sets A and B is defined to be $\inf d(U, V)$, where $U \in A$ and $V \in B$. When A and B are convex bodies, "sup" and "inf" can be replaced with "max" and "min."

Theorem 5.4. *Any two diameters of a finite point set meet at some point (either in the interior or at the endpoint).*

Theorem 5.5. *An Arbitrary intersection of convex sets is again convex.*

A supporting line of a convex set is a line which only intersects the boundary of the convex set. A supporting line divides the plane into two open half planes, one of which contains no points of the convex set.

Theorem 5.6. *If a bounded closed domain A covers a convex body B, then the perimeter of A is not less than the perimeter of B.*

Theorem 5.7. *There exists at least one supporting line passing through any given point on the boundary of a convex domain. If the supporting line is not unique for some boundary point, then there are infinitely many supporting lines through that point.*

Given a bounded convex domain and a line l, there exists a unique pair of supporting lines parallel to l. The convex domain lies in the strip bounded by these two parallel lines. The distance between these two parallel lines is called the width of the convex domain in the direction of l.

For a convex body, its maximum width is equal to its diameter. If two points in the convex body attain the diameter, the two lines passing through these points perpendicular to the line though them are two parallel supporting lines with a maximum width.

Separation of Convex Sets. If two convex domains have no common interior points, then there exists a separation line. In particular, if the distance between two convex domains is greater than 0, then there exists a separation line which does not intersect any of these domains.

Theorem 5.8. *The diameter of any planar finite point set is equal to the diameter of its convex hull.*

Theorem 5.9 (Helly's Theorem). *There are finitely many convex sets in a plane (or infinitely many convex bodies). If any three have a nonempty intersection, then all of them have a nonempty intersection.*

The assertion remains true if these convex sets are in the 3-dimensional Euclidean space and any four of them have a nonempty intersection.

(4) *Planar point sets*

Theorem 5.10 (Sylvester–Gallai Theorem). *Given $n(\geq 3)$ points in a plane. If every line passing through two of these points also passes a third one, then all points are collinear.*

This is a famous result that gave birth to a set of propositions. The following theorem follows from the Sylvester–Gallai theorem which is stronger.

Theorem 5.11. *Given $n(\geq 3)$ points in a plane, not all lying on a line. Then there are at least n different lines determined by these points, and at least one of these lines passes through exactly two points.*

The following theorem is a dual form of the Sylvest–Gallai theorem.

Theorem 5.12. *Given finitely many lines in a plane, such that no two are parallel. If for each two lines, there is a third line passing through their intersection, then all lines pass through a common point.*

By inversion, we also have the following variant on circles.

Theorem 5.13. *Given $n(\geq 4)$ points in a plane, such that no three are collinear. If every circle which passes through three of the points also passes through a fourth one, then all points are concyclic.*

Theorem 5.14 (Erdös-Szekeres Theorem). *For each integer $m \geq 3$, there exists a positive integer n_0 with the following property: for any n_0 points in a plane with no three collinear, there exist m points which are the vertices of a convex m-gon.*

(5) *Coverings and embeddings*

Covering is a well-known concept. Putting some shapes into some shape without overlapping is called embedding.

In the embedding of discs, we often use the "method of expansion." For example, to prove that a disc can be embedded into some region, the inner boundary is "shaved off" by one layer (with a thickness equal to the radius of the disc), and the problem boils down to whether there is still some point left after "shaving off." If there is one, it can be chosen as the center of the disc.

Theorem 5.15. *If a bounded region A covers B, then the area and diameter of A is not less than those of B. Furthermore, if B is convex, then the same is true for the perimeters.*

Theorem 5.16 (Overlapping Principle 1). *There are n segments d_i $(1 \leq i \leq n)$ placed in a unit segment d. If the total length of d_i is $>k$ (where k is a positive integer), then there exists a point in d which is covered by at least $k+1$ segments of d_i.*

Theorem 5.17 (Overlapping Principle 2). *There are n shapes S_i $(1 \leq i \leq n)$ placed in a shape S with area 1. If the total area of S_i is $>k$ (where k is a positive integer), then there exists a point in S which is covered by at least $k+1$ shapes of S_i.*

5.1.2   *Common methods*

Many combinatorial geometry problems are essentially existence problems or enumerative combinatorial problems, which can be solved using the methods of solving existence problems and enumerative problems in the previous chapters. We will not elaborate on them here. This chapter only introduces some specific methods to solve problems concerning covering and embedding.

(1)   *Intersection of large covers*

In covering problems, sometimes it is more complex to directly find the minimum covering. It is convenient to first find several large covers and then take their intersection.

Example 5.1. Prove that a finite point set F of diameter 1 can be covered by a closed disc of radius $\frac{\sqrt{3}}{2}$.

Proof. Since $d(F) = 1$, there exist two points $A, B \in F$ with $|AB| = 1$. For any $P \in F$, since $|PA| \leq 1$ and $|PB| \leq 1$, the point P lies in the two closed discs centered at A and B with radius 1, denoted by M_1 and M_2.

Let the boundary of the two discs intersect at C and D, and let the midpoint of CD be O. The disc M with center O and radius $|OC| = |OD| = \frac{\sqrt{3}}{2}$ covers F since $F \subseteq M_1 \cap M_2 \subseteq M$. Thus, F can be covered by a closed disc of radius $\frac{\sqrt{3}}{2}$.

(2) *Local to global*

Sometimes we can first treat some local properties of the graph, and then consider the global properties.

Example 5.2. Assume that a convex polygon M cannot cover any triangle with area 1. Show that M can be covered by a triangle with area 4.

Proof. Choose an inscribed triangle of M with a maximum area, say $\triangle ABC$. By assumption, we have $S_{\triangle ABC} < 1$. Draw lines though A, B, and C parallel to the opposite side, respectively, and form the triangle $A'B'C'$. Any point of M lies in the interior or on the boundary of $\triangle A'B'C'$ (otherwise it would contradict the maximality of $\triangle ABC$), that is, M is covered by $\triangle A'B'C'$. Note $S_{\triangle A'B'C'} = 4S_{\triangle ABC} < 4$, we may enlarge $\triangle A'B'C'$ a little bit to a triangle with area 4. Thus, M can be covered by a triangle with area 4.

(3) *Contraction and expansion*

It is sometimes useful to contract or expand the shapes for embedding problems.

Example 5.3. 16 non-overlapping discs with radius of 3 are placed in a circle with radius of 18. Prove that we can place another 9 non-overlapping discs with radius of 1 in the remaining part of the big circle.

Proof. First, reduce the radius 18 of the circle by 1, and increase the radius 3 of all 16 circles by 1. Since

$$\pi \cdot 17^2 - 16\pi \cdot 4^2 = 289\pi - 256\pi = 33\pi > 0,$$

there exists a point A_1 in the remaining part of the disc with radius 17 after 16 discs with radius 4 (possibly overlapping) are removed. Thus, the unit disc centered at A_1 lies in the original big circle with radius 18 and does not intersect the 16 discs with radius 3.

Assume we have embedded $k(1 \leq k \leq 8)$ unit discs without overlapping. Analogous to the above argument, since

$$\pi \cdot 17^2 - 16 \cdot \pi \cdot 4^2 - k \cdot \pi \cdot 2^2 = (33 - 4k)\pi > 0 \quad (1 \leq k \leq 8),$$

there exists some point A_{k+1} in the remaining part of the disc with radius 17 after 16 discs with radius 4 and k discs with radius 2 are removed. Thus, we can embed one more unit disc with center A_{k+1}. Finally, we can place up to 9 unit discs.

(4) *Coloring and valuation*

In some problems, it is possible to color or assign values to square grids and prove the assertion by contradiction.

Example 5.4. Can an 8×8 square grid be partitioned into 9 number of 2×2 squares like ⊞ and 7 *L*-shaped tetrominoes like ⌐?

Solution. Suppose it is possible. Color the 8×8 grid in the following way: the unit squares in the columns with odd indices black, and the squares in the columns with even indices white.

Since any 2×2 square covers exactly 2 black and 2 white unit squares, an *L*-shaped tetromino covers either 3 black and 1 white unit squares, or 1 black and 3 white unite squares. Since there are 7 *L*-shaped tetrominoes, we conclude that the number of black squares and the number of white squares do not coincide, which is clearly not the case.

Hence, such partition does not exist.

(5) *Translating the graphics*

In some questions, moving the graphics is very helpful in solving the problems.

Example 5.5. In the unit square are placed many discs with diameter 0.001, and the distance between the centers of any two discs is not less than 0.002. Prove that the total area covered by these discs is less than 0.34.

Proof. Let M be the set of given discs in the interior of the unit square, and let S be the area of the region covered by these discs. Let m be a direction parallel to some side of the square, and n be a direction rotated from m by $60°$. Translate M along the direction m by 0.001, denoted by M_1; also translate M along the direction n by 0.001, denoted by M_2. The area of M_1 and M_2 are both S.

Next, we claim that the discs in M, M_1, and M_2 are non-overlapping. In fact, if a disc $\odot O_1$ in M and a disc $\odot O_2'$ in M_1 have a common interior, then $O_1 O_2' < 0.001$. Assume that $\odot O_2'$ is translated from the disc $\odot O_2$ in M along the direction m by 0.001. Then, $O_2 O_2' = 0.001$ and consequently, $O_1 O_2 < O_1 O_2' + O_2 O_2' < 0.001 + 0.001 = 0.002$, which contradicts the assumption that the distance between any two centers of the discs in M is not less than 0.002.

Thus, any disc in M does not intersect any disc in M_1. Similarly, any disc in M does not intersect any disc in M_2. Since M_2 can be considered being translated from M_1 along direction l by a distance of 0.001 (as shown in Figure 5.1), we obtain similarly that any disc in M_2 does not intersect any disc in M_1. Since M, M_1, and M_2 lie in a square with side length 1.001, we see that $3S \leq (1.001)^2$, i.e., $S \leq \frac{1}{3} \times (1.001)^2 < 0.34$.

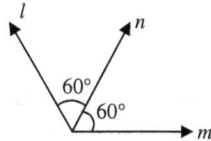

Figure 5.1

5.2 Problems and Solutions

5.2.1 *Point and line problems*

Problem 5.1 (IMO 6-5, proposed by Romania). Suppose five points in a plane are situated so that no two of the straight lines joining them are parallel, perpendicular, or coincident. From each point perpendiculars are drawn to all the lines joining the other four points. Determine the maximum number of intersections that these perpendiculars can have.

Solution. Let P_1, P_2, P_3, P_4, and P_5 be the five given points which determine $C_5^2 = 10$ lines. Since 4 points determine $C_4^2 = 6$ lines, 6 perpendicular lines are drawn from each point. Thus, there are a total of 30 perpendiculars with maximum $C_{30}^2 = 435$ intersections. However, as we shall now show, not all these intersections are distinct, and some are not even possible.

The three lines drawn from P_3, P_4, and P_5 perpendicular to P_1P_2 are parallel, so they do not intersect. This happens for each of the 10 lines with 3 parallel perpendiculars, and thus, $C_5^2 \cdot C_3^2 = 30$ intersections shall be deducted from the maximum number 435.

From each point of $P_1, P_2, \ldots, P_5$, the 6 number of lines are drawn perpendicular to the 6 lines determined by the other 4 points, and these six lines meet at a single point. Thus we need to subtract $5 \times (C_6^2 - 1) = 70$ points.

The original 5 points determine $C_5^3 = 10$ triangles. In each, the perpendiculars from the vertices to the opposite sides (the altitudes) are concurrent; they meet at one instead of 3 points. So, two points of the intersection are lost in each of the 10 triangles. Therefore, there are 20 fewer intersections, leaving $435 - 30 - 70 - 20 = 315$ possible intersections.

It is not hard to see that in general the lost intersections considered above fall into distinct categories and do not overlap. Hence, the maximum number of possible intersections is 315.

【Score Situation】 This particular problem saw the following distribution of scores among contestants: 3 contestants scored 7 points, no contestant scored 6 points, no contestant scored 5 points, 7 contestants scored 4 points, no contestant scored 3 points, 6 contestants scored 2 points, 1 contestant scored 1 point, and no contestant scored 0 point. The average score of this problem is 3.647, indicating that it was relatively straightforward.

Among the top five teams in the team scores, the scores of this problem are as follows: the Soviet Union team scored 35 points (with a total team score of 269 points), the Hungary team scored 37 points (with a total team score of 253 points), the Romania team scored 38 points (with a total team score of 213 points), the Poland team scored 33 points (with a total team score of 209 points), and the Bulgaria team scored 29 points (with a total team score of 198 points).

The gold medal cutoff for this IMO was set at 38 points (with 7 contestants earning gold medals), the silver medal cutoff was 31 points (with 9 contestants earning silver medals), and the bronze medal cutoff was 27 points (with 19 contestants earning bronze medals).

In this IMO, only one contestant achieved a perfect score of 42 points, namely David Bernstein from the Soviet Union.

Problem 5.2 (IMO 7-6, proposed by Poland). In a plane a set of n points ($n \geq 3$) is given. Each pair of points is connected by a segment. Let d be the length of the longest of these segments. We define a diameter of the set to be any connecting segment of length d. Prove that the number of diameters of the given set is at most n.

Proof. We shall prove the result by induction on n. For a set of $n = 3$ points, there are obviously at most three diameters. Assume $n \geq 4$ and the assertion holds for less than n points. Suppose on the contrary that a counter-example exists, namely there are more than n diameters.

If there is a point which is the endpoint of at most one diameter, we delete this point and there are at least n diameters in the configuration of the remaining $n - 1$ points, which contradicts the inductive hypothesis. Thus, each point is the endpoint of at least two diameters.

Since the number of diameters is larger than the number of points, there exists a point, say A, which is the endpoint of at least three diameters. Let AB, AC, and AD be three diameters with the common endpoint A. The angle between any two of these diameters is no larger than $60°$, for otherwise, the distance between two of B, C, and D would be larger than d.

Without loss of generality, we assume that AD lies between AB and AC. The common area of the three discs $\odot(A, d)$ (the closed disc with center A and radius d), $\odot(B, d)$ and $\odot(C, d)$ contains all the point set.

We show that the only point in this area at distance d to point D is the point A. In fact, as shown in Figure 5.2, for any point $G \neq A$ on arc $\overset{\frown}{AF}$, we have

$$\angle GDC > \angle ADC = \angle ACD > \angle GCD.$$

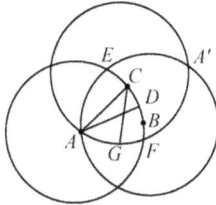

Figure 5.2

Hence $DG < CG = d$. This is also true for any point on arcs $\overset{\frown}{AE}$ and $\overset{\frown}{EF}$. Thus, there is only one point, namely A, at distance d to D, which contradicts the fact that each point is the endpoint of at least two diameters.

【Score Situation】 This particular problem saw the following distribution of scores among contestants: 10 contestants scored 9 points, no contestant scored 8 points, 2 contestants scored 7 points, no contestant scored 6 points, no contestant scored 5 points, 11 contestants scored 4 points, 13 contestants scored 3 points, 10 contestants scored 2 points, 13 contestants scored 1 point, and 21 contestants scored 0 point. The average score of this problem is 2.750, indicating that it had a certain level of difficulty.

Among the top five teams in the team scores, the scores of this problem are as follows: the Soviet Union team scored 61 points (with a total team score of 281 points), the Hungary team scored 46 points (with a total team score of 244 points), the Romania team scored 21 points (with a total team score of 222 points), the Poland team scored 8 points (with a total team score of 178 points), and the German Democratic Republic team scored 22 points (with a total team score of 175 points).

The gold medal cutoff for this IMO was set at 38 points (with 8 contestants earning gold medals), the silver medal cutoff was 30 points (with 12 contestants earning silver medals), and the bronze medal cutoff was 20 points (with 17 contestants earning bronze medals).

In this IMO, only two contestants achieved a perfect score of 40 points, namely László Lovász from Hungary and Pavel Bleher from the Soviet Union.

Problem 5.3 (IMO 11-5, proposed by Mongolia). Given $n > 4$ points in a plane such that no three are collinear. Prove that there are at least C_{n-3}^2 convex quadrilaterals whose vertices are four of the given points.

Proof. Consider first the case $n = 5$. If the convex hull of the five points has four of them on its boundary, then they form a convex quadrilateral. If the boundary of the convex hull contains only three of the points, say A, B, and C, then the other two points, D and E, are inside triangle ABC.

Assume without loss of generality that the line DE does not intersect the side AB, as in Figure 5.3. Then $ABDE$ is a convex quadrilateral. Thus, the assertion holds for $n = 5$.

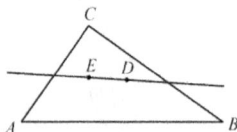

Figure 5.3

Consider now the general case $n \geq 5$. With each of the C_n^5 subsets of five of the n points, associate one of the convex quadrilaterals whose existence was demonstrated above.

Each quadrilateral is associated with at most $n - 4$ quintuples of points. Therefore, there are at least

$$\frac{C_n^5}{n-4} = \frac{n(n-1)(n-2)(n-3)}{5!}$$

different convex quadrilaterals in the give set of n points.

It remains to prove

$$\frac{n(n-1)(n-2)(n-3)}{120} \geq C_{n-3}^2,$$

that is

$$n(n-1)(n-2) \geq 60(n-4). \tag{1}$$

For $n = 6, 7, 8$, we have $6 \times 5 \times 4 \geq 60 \times 2$, $7 \times 6 \times 5 \geq 60 \times 3$, and $8 \times 7 \times 6 \geq 60 \times 4$, so (1) holds in these cases. For $n \geq 9$, we have

$$n(n-1)(n-2) \geq 9 \times 8 \times (n-4) > 60(n-4),$$

(1) also holds. This completes the proof.

【Score Situation】 This particular problem saw the following distribution of scores among contestants: 38 contestants scored 7 points, 10 contestants scored 6 points, 10 contestants scored 5 points, 10 contestants scored 4 points, 9 contestants scored 3 points, 7 contestants scored 2 points, 11 contestants scored 1 point, and 17 contestants scored 0 point. The average score of this problem is 4.179, indicating that it was simple.

Among the top five teams in the team scores, the scores of this problem are as follows: the Hungary team scored 48 points (with a total team score of 247 points), the German Democratic Republic team scored 49 points (with a total team score of 240 points), the Soviet Union team scored 48 points (with a total team score of 231 points), the Romania team scored 47 points (with a total team score of 219 points), and the United Kingdom team scored 41 points (with a total team score of 193 points).

The gold medal cutoff for this IMO was set at 40 points (with 3 contestants earning gold medals), the silver medal cutoff was 30 points (with 20 contestants earning silver medals), and the bronze medal cutoff was 24 points (with 21 contestants earning bronze medals).

In this IMO, only three contestants achieved a perfect score of 40 points, namely Tibor Fiala from Hungary, Vladimir Drinfeld from the Soviet Union, and Simon Phillips Norton from the United Kingdom.

Problem 5.4 (IMO 13-5, proposed by Bulgaria). Prove that for every natural number m, there exists a finite set S of points in a plane with the following property: For every point A in S, there are exactly m points in S which are at unit distance from A.

Proof. We induct on m.

For $m = 1$, the two endpoints of a unit segment form the required set S.

Assume that the assertion holds for $m = k$, i.e., there exists a set S_k such that for every point A in S_k, there are exactly k points in S_k which are at unit distance from A.

We translate S_k with respect to a unit vector $\vec{v}$, and denote the resulting set by S_k'. The vector $\vec{v}$ is required to satisfy the following two conditions:

(1) $\vec{v}$ is not parallel to any line determined by any two points in S_k;
(2) For each point in S_k, construct a circle centered at this point with radius 1. Mark the intersection points of these unit circles. We require that $\vec{v}$ is not parallel to any line through a marked point and the center of the circles it lies.

Let $S_{k+1} = S_k \cup S_k'$. By (1), we see that S_k and S_k' do not intersect. For every point A in S_{k+1}, there are exactly $k+1$ points in S_{k+1} which are at unit distance from A.

In fact, if $A \in S_k$, then there are exactly k points in S_k which are at unit distance from A. The point $A + \vec{v}$ in S'_k is at unit distance from A. Condition (2) guarantees that no other point $B + \vec{v}$ is at unit distance from A (for otherwise $C = B + \vec{v}$ is the intersection point of unit circles centered at A and B, so $\vec{v}$ is parallel to BC, which contradicts condition (2)).

Thus there are exactly $k + 1$ points in S_{k+1} which are at unit distance from A. A similar argument works for $A \in S'_k$. Therefore, the assertion is true for $m = k + 1$. Hence, by induction, the assertion is true for every positive integer m.

【Score Situation】 This particular problem saw the following distribution of scores among contestants: 4 contestants scored 7 points, no contestant scored 6 points, no contestant scored 5 points, no contestant scored 4 points, 1 contestant scored 3 points, 2 contestants scored 2 points, 4 contestants scored 1 point, and 17 contestants scored 0 point. The average score of this problem is 1.393, indicating that it was relatively challenging.

Among the top five teams in the team scores, the scores of this problem are as follows: the Hungary team scored 46 points (with a total team score of 255 points), the Soviet Union team scored 51 points (with a total team score of 205 points), the German Democratic Republic team scored 28 points (with a total team score of 142 points), the Poland team scored 28 points (with a total team score of 118 points), the Romania team scored 16 points (with a total team score of 110 points), and the United Kingdom team scored 12 points (with a total team score of 110 points).

The gold medal cutoff for this IMO was set at 35 points (with 7 contestants earning gold medals), the silver medal cutoff was 23 points (with 12 contestants earning silver medals), and the bronze medal cutoff was 11 points (with 29 contestants earning bronze medals).

In this IMO, only one contestant achieved a perfect score of 42 points, namely Imre Ruzsa from Hungary.

Problem 5.5 (IMO 17-5, proposed by the Soviet Union). Determine, with a proof, whether or not one can find 1975 points on the circumference of a circle with unit radius such that the distance between any two of them is a rational number.

Solution. Let $\theta_k = \arctan \frac{k^2 - 1}{2k}$ for $1 \leq k \leq 1975$. We shall show that the 1975 points $P_k(\cos 2\theta_k, \sin 2\theta_k)$ $(1 \leq k \leq 1975)$ satisfy the prescribed property.

Indeed, since $\tan \theta_k = \frac{k^2 - 1}{2k}$,

$$\sin \theta_k = \frac{k^2 - 1}{k^2 + 1} \quad \text{and} \quad \cos \theta_k = \frac{2k}{k^2 + 1}.$$

For $1 \leq i < j \leq 1975$, a straightforward calculation gives

$$
\begin{aligned}
|P_i P_j| &= \sqrt{(\cos 2\theta_j - \cos 2\theta_i)^2 + (\sin 2\theta_j - \sin 2\theta_i)^2} \\
&= \sqrt{2 - 2\cos 2(\theta_j - \theta_i)} \\
&= 2\sin(\theta_j - \theta_i) \\
&= 2(\sin\theta_j \cos\theta_i - \cos\theta_j \sin\theta_i) \\
&= \frac{4(j - i)(ij - 1)}{(i^2 + 1)(j^2 + 1)}.
\end{aligned}
$$

Therefore, $|P_i P_j|$ is a positive rational number.

In conclusion, we have shown that $P_1, P_2, \ldots, P_{1975}$ are 1975 distinct points on the unit circle and the distance between any two of them is a rational number.

【Score Situation】This particular problem saw the following distribution of scores among contestants: 49 contestants scored 6 points, 4 contestants scored 5 points, 1 contestant scored 4 points, 4 contestants scored 3 points, 9 contestants scored 2 points, 12 contestants scored 1 point, and 49 contestants scored 0 point. The average score of this problem is 2.813, indicating that it had a certain level of difficulty.

Among the top five teams in the team scores, the scores of this problem are as follows: the Hungary team scored 38 points (with a total team score of 258 points), the German Democratic Republic team scored 30 points (with a total team score of 249 points), the United States team scored 34 points (with a total team score of 247 points), the Soviet Union team scored 42 points (with a total team score of 246 points), and the United Kingdom team scored 30 points (with a total team score of 241 points).

The gold medal cutoff for this IMO was set at 38 points (with 8 contestants earning gold medals), the silver medal cutoff was 32 points (with 25 contestants earning silver medals), and the bronze medal cutoff was 23 points (with 36 contestants earning bronze medals).

In this IMO, a total of six contestants achieved a perfect score of 40 points.

Problem 5.6 (IMO 23-6, proposed by Vietnam). Let S be a square with sides of length 100, and let L be a path within S which does not meet itself and which is composed of line segments $A_0 A_1, A_1 A_2, \ldots, A_{n-1} A_n$ with $A_0 \neq A_n$. Suppose that for every point P of the boundary of S there is a point of L at a distance from P not greater than $\frac{1}{2}$.

Prove that there are two points X and Y in L such that the distance between X and Y is not greater than 1, and the length of that part of L which lies between X and Y is not smaller than 198.

Proof. Denote the distance between a point P on the side of the square and a point Q on the polygonal path L by $d(P, Q)$, and denote the length of the polygonal path from point A to point B by $s(A, B)$. When

$$s(A_0, A) < s(A_0, B),$$

we shall write $A < B$. Let S_1, S_2, S_3, and S_4 be the vertices of the square. On L, take points S_1', S_2', S_3', and S_4' such that $d(S_i, S_i') \leq \frac{1}{2}$, as shown in Figure 5.4. We may name the points so that $S_1' < S_4' < S_2'$.

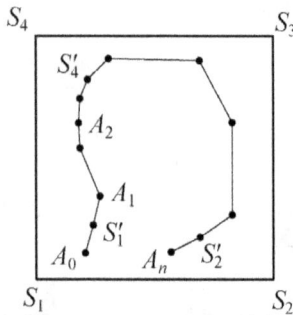

Figure 5.4

Now let L_1 be the set of all X on L such that $X \leq S_4'$, and let L_2 be the set of all X on L such that $X \geq S_4'$. Consider side $S_1 S_2$. There is a subset L_1' of $S_1 S_2$ whose points are of distance $\leq \frac{1}{2}$ from L_1 and a subset L_2' of $S_1 S_2$ whose points are of distance $\leq \frac{1}{2}$ from L_2. (Since L_1' includes S_1 and L_2' includes S_2, neither set is empty.)

The union of L_1' and L_2' is the side $S_1 S_2$, and because of the condition on the distances, the intersection of L_1' and L_2' is not empty. Let M be a point common to L_1' and L_2'.

Now select a point X in L_1 and a point Y in L_2 such that

$$d(M, X) \leq \frac{1}{2} \quad \text{and} \quad d(M, Y) \leq \frac{1}{2}.$$

Then $d(X, Y) \leq 1$. Also $X < S_4' < Y$ and

$$s(X, Y) = s(X, S_4') + s(S_4', Y) \geq 99 + 99 = 198.$$

【Score Situation】 This particular problem saw the following distribution of scores among contestants: 20 contestants scored 7 points, 6 contestants scored 6 points, 7 contestants scored 5 points, 8 contestants scored 4 points, 9 contestants scored 3 points, 11 contestants scored 2 points, 12 contestants scored 1 point, and 46 contestants scored 0 point. The average score of this problem is 2.555, indicating that it had a certain level of difficulty.

Among the top five teams in the team scores, the scores of this problem are as follows: the Germany team scored 25 points (with a total team score of 145 points), the Soviet Union team scored 21 points (with a total team score of 137 points), the United States team scored 24 points (with a total team score of 136 points), the German Democratic Republic team scored 8 points (with a total team score of 136 points), and the Vietnam team scored 19 points (with a total team score of 133 points).

The gold medal cutoff for this IMO was set at 37 points (with 10 contestants earning gold medals), the silver medal cutoff was 30 points (with 20 contestants earning silver medals), and the bronze medal cutoff was 21 points (with 31 contestants earning bronze medals).

In this IMO, only three contestants achieved a perfect score of 42 points, namely Bruno Haible from Germany, Grigori Perelman from the Soviet Union, and Lê Tự Quốc Thắng from Vietnam.

Problem 5.7 (IMO 24-4, proposed by Belgium). Let $\triangle ABC$ be an equilateral triangle and E the set of all points contained in the three segments AB, BC, and CA (including A, B, and C).

Determine whether, for every partition of E into two disjoint subsets, at least one of the two subsets contains the vertices of a right-angled triangle. Justify your answer.

Solution. The answer is in the affirmative.

Choose points P, Q, and R on BC, CA, and AB respectively, such that

$$AR : RB = BP : PC = CQ : QA = 1 : 2.$$

It follows easily that $PQ \perp CA$, $QR \perp AB$, and $RP \perp BC$, as shown in Figure 5.5.

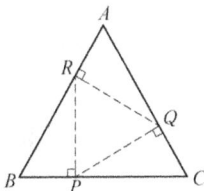

Figure 5.5

Consider an arbitrary two-coloring of the points of E, say red and blue. Two of the points P, Q, and R must have the same color; say P and Q are red. If there is another point X on side AC which is colored red, then $\triangle PQX$ is a right-angled triangle whose vertices are all red.

Assume now that all points on side AC other than Q are colored blue. If there is a point $Y \neq A$ on side AB which is colored blue, let Z be a point on AC such that $YZ \perp AC$. Then Z is blue and the right-angled triangle AYZ has three blue vertices.

Assume now that all points on side AB other than A are colored red. Then $\triangle BPR$ is a right-angled triangle whose vertices are all red.

The proof is completed.

【Score Situation】 This particular problem saw the following distribution of scores among contestants: 20 contestants scored 7 points, no contestant scored 6 points, no contestant scored 5 points, 2 contestants scored 4 points, 3 contestants scored 3 points, 2 contestants scored 2 points, 1 contestant scored 1 point, and 12 contestants scored 0 point. The average score of this problem is 4.050, indicating that it was simple.

Among the top five teams in the team scores, the Germany team achieved a total score of 212 points, the United States team achieved a total score of 171 points, the Hungary team achieved a total score of 170 points, the Soviet Union team achieved a total score of 169 points, and the Romania team achieved a total score of 161 points.

The gold medal cutoff for this IMO was set at 38 points (with 9 contestants earning gold medals), the silver medal cutoff was 26 points (with 27 contestants earning silver medals), and the bronze medal cutoff was 15 points (with 57 contestants earning bronze medals).

In this IMO, a total of four contestants achieved a perfect score of 42 points.

Problem 5.8 (IMO 25-3, proposed by Romania). In a plane two different points O and A are given. For each point X of the plane, other than O, denote by $a(X)$ the measure of the angle between OA and OX in radians, counterclockwise from $OX (0 \leq a(X) < 2\pi)$. Let $C(X)$ be the circle with center O and radius of length $OX + \frac{a(X)}{OX}$. Each point of the plane is colored by one of a finite number of colors.

Prove that there exists a point Y for which $a(Y) > 0$ such that its color appears on the circumference of the circle $C(Y)$.

Proof. Suppose every point of the plane has been colored with one of n different colors. For each positive r, let C_r be the circle with center O and radius r. Since there are only $2^n - 1$ different color combinations that may appear on C_r, there exist $0 < r < s < 1$ such that C_r and C_s bear the same set of colors.

We claim that there is a point Y on C_r such that the circle $C(Y)$, whose radius is

$$r + \frac{a(Y)}{r},$$

coincides with C_s; that is, $r + \frac{a(Y)}{r} = s$, or $a(Y) = r(s - r) \in (0, 1)$.

Clearly there is a point $Y \in C_r$ with $a(Y) = r(s - r)$. Since the circles C_r and C_s bear the same set of colors and $C(Y) = C_s$, the color of Y appears on the circle $C(Y)$.

【Score Situation】This particular problem saw the following distribution of scores among contestants: 29 contestants scored 7 points, 6 contestants scored 6 points, 3 contestants scored 5 points, 1 contestant scored 4 points, no contestant scored 3 points, 14 contestants scored 2 points, 18 contestants scored 1 point, and 121 contestants scored 0 point. The average score of this problem is 1.583, indicating that it was relatively challenging.

Among the top five teams in the team scores, the scores of this problem are as follows: the Soviet Union team scored 34 points (with a total team score of 235 points), the Bulgaria team scored 20 points (with a total team score of 203 points), the Romania team scored 21 points (with a total team score of 199 points), the United States team scored 35 points (with a total team score of 195 points), and the Hungary team scored 22 points (with a total team score of 195 points).

The gold medal cutoff for this IMO was set at 40 points (with 14 contestants earning gold medals), the silver medal cutoff was 26 points (with 35 contestants earning silver medals), and the bronze medal cutoff was 17 points (with 49 contestants earning bronze medals).

In this IMO, a total of eight contestants achieved a perfect score of 42 points.

Problem 5.9 (IMO 28-5, proposed by the German Democratic Republic). Let n be an integer greater than or equal to 3. Prove that there is a set of n points in a plane such that the distance between any two points is irrational and each set of three points determines a non-degenerate triangle with a rational area.

Proof 1. As shown in Figure 5.6, on the semi-circle which centered at O we select n points P_i $(i = 1, 2, \ldots, n)$ counter-clockwise. Let $\varphi_i = \angle P_i O P_{i+1}$ which is to be determined later, and denote the radius of this circle by r.

To calculate the length of the segment $P_i P_{i+k}$, by the law of sine,

$$d_{i,i+k} = 2r \cdot \sin\left(\sum_{j=i}^{i+k-1} \frac{\varphi_j}{2}\right) = r D_{i,i+k}. \tag{1}$$

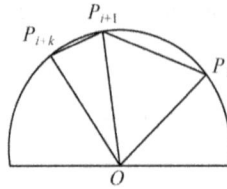

Figure 5.6

By trigonometric formulas, if all $\sin \frac{\varphi_i}{2}$ and $\cos \frac{\varphi_i}{2}$ are rational, then

$$D_{i,i+k} = 2 \sin \left(\sum_{j=i}^{i+k-1} \frac{\varphi_i}{2} \right) \tag{2}$$

is also rational. Since

$$\left(\frac{a^2 - 1}{a^2 + 1} \right)^2 + \left(\frac{2a}{a^2 + 1} \right)^2 = 1, \tag{3}$$

$$\lim_{a \to \infty} \frac{2a}{a^2 + 1} = 0 \tag{4}$$

for any positive $\delta > 0$, there exists a positive rational number a, such that

$$0 < \frac{2a}{a^2 + 1} < \delta. \tag{5}$$

There exists $x \in (0, \frac{\pi}{2})$ such that $\sin x = \frac{2a}{a^2+1}$. Now choose $\delta = \sin \frac{\pi}{2(n+1)}$, a positive rational a satisfying (5), and $x \in (0, \frac{\pi}{2})$ satisfying $\sin x = \frac{2a}{a^2+1}$. Then $x \in (0, \frac{\pi}{2(n+1)})$. We obtain from (3) that $\cos x$ is also rational.

Finally, take $\frac{\varphi_1}{2} = \frac{\varphi_2}{2} = \cdots = \frac{\varphi_n}{2} = x$. Since

$$\sum_{i=1}^{n} \frac{\varphi_i}{2} = nx = n\frac{\pi}{2n+1} < \frac{\pi}{2},$$

we have $\sum_{i=1}^{n} \varphi_i < \pi$, so the n points lie on the same semi-circle.

By (2) and the choice of φ_i, we see that $D_{i,i+k}$ is rational. Take $r = \sqrt{2}$, and then by (1), $d_{i,i+k} = \sqrt{2}D_{i,i+k}$ is irrational.

Let the area of $\triangle OP_i P_{i+k}$ be $S_{i,i+k}$. Then

$$S_{i,i+k} = \frac{1}{2} |OP_i| |OP_{i+k}| \sin \angle P_i OP_{i+k}$$

$$= \frac{1}{2}\sqrt{2} \cdot \sqrt{2} \sin \left(\sum_{j=i}^{i+k-1} \varphi_j \right) = \sin \left(\sum_{j=i}^{i+k-1} \varphi_j \right),$$

which is rational. For any three points $P_i, P_{i+j}, P_{i+k}(j < k)$, the area of $\triangle P_i P_{i+j} P_{i+k}$ is

$$S = S_{i,i+j} + S_{i+j,i+k} - S_{i,i+k},$$

which is rational. Thus, the assertion is proved.

Proof 2. We choose points $\{P_i\}$ on the graph of parabola $y = x^2$:

$$P_i(i, i^2), \quad i = 1, 2, \ldots, n.$$

Since any line intersects a parabola at no more than two points, no three points of $P_1, P_2, \ldots, P_n$ are collinear.

For any two points $P_i, P_j (i \neq j)$, the distance between them is given by

$$|P_i P_j| = \sqrt{(i-j)^2 + (i^2 - j^2)^2} = |i - j|\sqrt{1 + (i+j)^2},$$

which is clearly an irrational number.

For any three points P_i, P_j, P_k $(i < j < k)$, the area of $\triangle P_i P_j P_k$ is given by

$$S = \frac{1}{2} \begin{vmatrix} 1 & 1 & 1 \\ i & j & k \\ i^2 & j^2 & k^2 \end{vmatrix},$$

which is clearly a rational number. Thus, the assertion is proved.

【Score Situation】 This particular problem saw the following distribution of scores among contestants: 119 contestants scored 7 points, 6 contestants scored 6 points, 10 contestants scored 5 points, 8 contestants scored 4 points, 7 contestants scored 3 points, 7 contestants scored 2 points, 21 contestants scored 1 point, and 59 contestants scored 0 point. The average score of this problem is 4.249, indicating that it was simple.

Among the top five teams in the team scores, the scores of this problem are as follows: the Romania team scored 42 points (with a total team score of 250 points), the Germany team scored 42 points (with a total team score of 248 points), the Soviet Union team scored 42 points (with a total team score of 235 points), the German Democratic Republic team scored 42 points (with a total team score of 231 points), and the United States team scored 36 points (with a total team score of 220 points).

The gold medal cutoff for this IMO was set at 42 points (with 22 contestants earning gold medals), the silver medal cutoff was 32 points (with 42 contestants earning silver medals), and the bronze medal cutoff was 18 points (with 56 contestants earning bronze medals).

In this IMO, a total of 22 contestants achieved a perfect score of 42 points.

Problem 5.10 (IMO 36-3, proposed by Japan). Determine all integers $n > 3$ for which there exist n points $A_1, A_2, \ldots, A_n$ in a plane, no three collinear, and real numbers $r_1, r_2, \ldots, r_n$ such that for $1 \le i < j < k \le n$, the area of $\triangle A_i A_j A_k$ is $r_i + r_j + r_k$.

Solution. The only integer with the required property is $n = 4$.

Suppose a configuration of points $A_1, A_2, \ldots, A_n$ satisfies the required property. We first prove the following two lemmas.

Lemma 1. *For any triangle, say $\triangle A_1 A_2 A_3$, there do not exist two points, say A_4 and A_5, such that $A_1 A_2 A_3 A_4$ and $A_1 A_2 A_3 A_5$ are both convex (vertices are listed in this order).*

For otherwise, by assumption, we have

$$S_{\triangle A_1 A_2 A_3} + S_{\triangle A_1 A_3 A_4} = S_{A_1 A_2 A_3 A_4} = S_{\triangle A_1 A_2 A_4} + S_{\triangle A_2 A_3 A_4}.$$

Therefore, $r_1 + r_2 + r_3 + r_1 + r_3 + r_4 = r_1 + r_2 + r_4 + r_2 + r_3 + r_4$, and hence $r_1 + r_3 = r_2 + r_4$. Similarly, $r_1 + r_3 = r_2 + r_5$. We obtain $r_4 = r_5$, which implies that $S_{\triangle A_1 A_2 A_4} = S_{\triangle A_1 A_2 A_5}$.

Consequently $A_4 A_5 \parallel A_1 A_2$. Similarly, $A_4 A_5 \parallel A_2 A_3$, which implies that A_1, A_2, and A_3 are collinear, a contradiction.

Lemma 2. *For any triangle, say $\triangle A_1 A_2 A_3$, there do not exist two points in the interior, say A_4 and A_5.*

For otherwise, from

$$S_{\triangle A_1 A_2 A_3} = S_{\triangle A_1 A_2 A_4} + S_{\triangle A_1 A_3 A_4} + S_{\triangle A_2 A_3 A_4},$$

we obtain

$$r_1 + r_2 + r_3 = r_1 + r_2 + r_4 + r_1 + r_3 + r_4 + r_2 + r_3 + r_4.$$

Thus $r_4 = -\frac{r_1 + r_2 + r_3}{3}$. Similarly, $r_5 = -\frac{r_1 + r_2 + r_3}{3}$. Therefore, $r_4 = r_5$. It follows from the same argument as in Lemma 1 that A_1, A_2, and A_3 are collinear, which is a contradiction.

Suppose $n > 4$. Consider the convex hull of $A_1, A_2, \ldots, A_n$. If there are more than 4 points on the boundary, then we have a convex pentagon, say $A_1 A_2 A_3 A_4 A_5$. Then $A_1 A_2 A_3 A_4$ and $A_1 A_2 A_3 A_5$ are convex quadrilaterals, which is impossible by Lemma 1.

If the convex hull is a quadrilateral, say $A_1 A_2 A_3 A_4$, and A_5 is in the interior of $A_1 A_2 A_3 A_4$. Without loss of generality, we may assume that A_5 is in the interior of $\triangle A_1 A_3 A_4$. Then $A_1 A_2 A_3 A_4$ and $A_1 A_2 A_3 A_5$ are convex quadrilaterals, which is impossible. If the convex hull is a triangle, then it is impossible by Lemma 2.

Thus, $n \leq 4$. It is indeed possible for $n = 4$. Let $A_1 A_2 A_3 A_4$ be a unit square and $r_1 = r_2 = r_3 = r_4 = \frac{1}{6}$. Then

$$S_{\triangle A_i A_j A_k} = \frac{1}{2} = r_i + r_j + r_k \quad (1 \leq i < j < k \leq 4).$$

We have showed that $n = 4$ is the only integer with the required property.

【Score Situation】 This particular problem saw the following distribution of scores among contestants: 100 contestants scored 7 points, 17 contestants scored 6 points, 24 contestants scored 5 points, 34 contestants scored 4 points, 28 contestants scored 3 points, 33 contestants scored 2 points, 80 contestants scored 1 point, and 96 contestants scored 0 point. The average score of this problem is 3.126, indicating that it was relatively straightforward.

Among the top five teams in the team scores, the scores of this problem are as follows: the China team scored 40 points (with a total team score of 236 points), the Romania team scored 42 points (with a total team score of 230 points), the Russia team scored 42 points (with a total team score of 227 points), the Vietnam team scored 40 points (with a total team score of 220 points), and the Hungary team scored 40 points (with a total team score of 210 points).

The gold medal cutoff for this IMO was set at 37 points (with 30 contestants earning gold medals), the silver medal cutoff was 29 points (with 71 contestants earning silver medals), and the bronze medal cutoff was 19 points (with 100 contestants earning bronze medals).

In this IMO, a total of 14 contestants achieved a perfect score of 42 points.

Problem 5.11 (IMO 40-1, proposed by Estonia). Determine all finite sets S of at least three points in a plane which satisfy the following condition: for any two distinct points A and B in S, the perpendicular bisector of the line segment AB is an axis of symmetry for S.

Solution 1. Observe that the set of all vertices of a regular polygon satisfies the prescribed property. We now show that these are the only solutions.

Suppose that S satisfies the given condition. Let T be the convex hull of S. It is easy to see that T is not a segment. Hence we assume T is a convex polygon $P_1 P_2 \cdots P_n$.

If there is a point in S other than the vertices of $P_1 P_2 \cdots P_n$, say Q, then assume without loss of generality that

$$QP_1 = \max\{QP_1, QP_2, \ldots, QP_n\} = \max_{P \in T} QP.$$

Let l be the perpendicular bisector of QP_1. Then S is symmetric with respect to l by assumption. Let P_i be a vertex on the same side of l as Q

such that the distance to l is maximal. Then $d(P_i, l) > d(Q, l)$, where we use $d(P, l)$ to denote the distance from a point P to line l. The symmetric point P'_i of P_i with respect to l lies on the same side of l as P_1, and

$$d(P'_i, l) = d(P_i, l) > d(Q, l) = d(P_1, l).$$

Hence P'_i lies outside T, which contradicts the condition. Thus, S is the set of all the vertices of $P_1 P_2 \cdots P_n$.

Consider the perpendicular bisector of $P_1 P_3$, which is a symmetric axis of S. We conclude that P_2 lies on this axis since it is the only point on its side of line $P_1 P_3$. Therefore, $P_1 P_2 = P_2 P_3$. Similarly,

$$P_1 P_2 = P_2 P_3 = P_3 P_4 = \cdots = P_1 P_n.$$

For $n = 3$, we see that S is the set of the three vertices of a regular triangle.

For $n \geq 4$, consider the perpendicular bisector of $P_1 P_4$ which is a symmetric axis of S. We see that P_2 and P_3 are symmetric about this axis. Thus $\angle P_1 P_2 P_3 = \angle P_2 P_3 P_4$. Similarly,

$$\angle P_1 P_2 P_3 = \angle P_2 P_3 P_4 = \cdots = \angle P_n P_1 P_2.$$

Therefore, $P_1 P_2 \cdots P_n$ is a regular polygon and S is the set of all the vertices of a regular polygon.

Solution 2. Let G be the barycenter of S. For any two distinct points A and B in S, denote the map of symmetry with respect to the perpendicular bisector of AB by r_{AB}. Since $r_{AB}(S) = S$, by assumption, $r_{AB}(G) = G$, which implies that the distance from any point in S to G is the same. Thus, the points in S lie on a circle with center G. In particular, the points in S are the vertices of a convex polygon, say $P_1 P_2 \cdots P_n$ $(n \geq 3)$.

Consider the map $r_{P_1 P_3}$, and the same argument in the first solution shows that

$$r_{P_1 P_3}(P_2) = P_2.$$

Hence $P_1 P_2 = P_2 P_3$. Similarly,

$$P_2 P_3 = P_3 P_4 = \cdots = P_n P_1.$$

Thus, $P_1 P_2 \cdots P_n$ is a regular n-gon. It is easy to verify that the set of all the vertices of a regular polygon has the prescribed property. Thus, the required finite set S is the set of all vertices of a regular polygon.

【Score Situation】 This particular problem saw the following distribution of scores among contestants: 140 contestants scored 7 points, 48 contestants scored 6 points, 41 contestants scored 5 points, 42 contestants scored 4 points, 45 contestants scored 3 points, 53 contestants scored 2 points, 52 contestants scored 1 point, and 29 contestants scored 0 point. The average score of this problem is 4.298, indicating that it was simple.

Among the top five teams in the team scores, the scores of this problem are as follows: the Russia team scored 42 points (with a total team score of 182 points), the China team scored 40 points (with a total team score of 182 points), the Vietnam team scored 41 points (with a total team score of 177 points), the Romania team scored 35 points (with a total team score of 173 points), and the Bulgaria team scored 41 points (with a total team score of 170 points).

The gold medal cutoff for this IMO was set at 28 points (with 38 contestants earning gold medals), the silver medal cutoff was 19 points (with 70 contestants earning silver medals), and the bronze medal cutoff was 12 points (with 118 contestants earning bronze medals).

In this IMO, no contestant achieved a perfect score of 42 points.

Problem 5.12 (IMO 52-2, proposed by Britain). Let S be a finite set of at least two points in a plane. Assume that no three points of S are collinear. A *windmill* is a process that starts with a line l going through a single point $P \in S$. The line rotates clockwise about the *pivot* P until the first time that the line meets some other point belonging to S. This point, Q, takes over as the new pivot, and the line now rotates clockwise about Q, until it next meets a point of S. This process continues indefinitely.

Show that we can choose a point P in S and a line l going through P such that the resulting windmill uses each point of S as a pivot infinitely many times.

Proof. Consider each line l as having a direction, which varies continuously when the line l rotates. First, consider the case when $|S| = 2k + 1$ is odd.

In the process of a windmill, we say a line l is on a *good* position. if there are k points of S on each side of the line l when l just leaves one point of S.

We consider two good positions to be equal if l passes the same two points with the same direction; otherwise they are different good positions. It is easy to see that the number of good positions is finite in all windmills. Given a positive direction such as x-axis, denote all good positions in the order of clockwise angles in $[0, 2\pi)$ by $l_1, l_2, \ldots, l_m$. We proceed to prove it in three steps.

Firstly, given any direction, we can see that there is at most one good position. Since if there were two good positions l_i and l_j that had the same direction, then they were parallel and did not coincide, and the number on the right side to the lines l_i and l_j should be k or $k-1$, which could not be possible to both lines l_i and l_j.

Secondly, for any point in S, there exists some l_i passing through P as follows.

Take any point $P \in S$ and line l passing through P but not through the other point of S. If the number of points to the right side of line l is s, then the number of points to the left side of line l is $2k - s$. The difference of two numbers is $2k - 2s$.

Now, rotate l about P clockwise, with s increasing or decreasing by 1 when l is passing a point, so that the number $2k - 2s$ is changed by 2. When l rotates $180°$, the number $2k - 2s$ becomes its opposite number. Therefore, there exists a moment for an l when the numbers of points on two sides are equal. Denote the last passing point by Q. Then there is a good position l_i passing P and Q.

Thirdly, a windmill starting from a good position meets the requirement of the problem.

Without loss of generality, we may start with l_1. And we show that the next meeting point is just l_2, so the windmill meets all l_1 and continues infinitely many times. By step two, we know that each point of S is used as a pivot infinitely many times.

Suppose that $l = l_1$ when passing P and Q with P as a pivot and dividing S equally when l_1 is just leaving Q. Suppose that l meets point R next and use R as a pivot. Then l remains to divide S equally when leaving P, which is valid for R on any side of P.

This shows that l is still a good position when it meets R, and denote this good position as l'. It remains to be shown that there is no good position between l and l', so $l' = l_2$. Since if l_2 was a good position between l and l', take directed line l'' passing P and parallel to l_2, then l'' divides S equally, so there are at least $k + 1$ points on the left or right side of l_2, which contradicts the fact that l_2 is a good position.

Now, consider the case when $|S| = 2k$ is even. The argument above is still valid with a suitable revision. We say a line l is in a good position if there are k points of S on the right side of the line l when l just leaves one point of S. Then the first and second steps can be proven similarly. For the third step, starting from a *good* position, we can show similarly that the next meeting point by l is still a good position. The only slight difference to the odd case is to show that there is no good position between l_1 and l_2.

Take a directed line l'' passing P and parallel to l_2. Then there are k points of S to the right of l''. If l_2 were on the right side to l'', then there would be no more than $k - 2$ points, so l'' would not be a good position. If l_2 were on the left side to l'', then there would be at least $k + 1$ to the right side to l'', so l'' would not be a good position. Thus, we have proved it for the even case.

(Based on the solution by Zhou Tianyou.)

【Score Situation】 This particular problem saw the following distribution of scores among contestants: 22 contestants scored 7 points, 5 contestants scored 6 points, 2 contestants scored 5 points, 4 contestants scored 4 points, 2 contestants scored 3 points, 14 contestants scored 2 points, 124 contestants scored 1 point, and 390 contestants scored 0 point. The average score of this problem is 0.654, indicating that it was extremely difficult.

Among the top five teams in the team scores, the scores of this problem are as follows: the China team scored 12 points (with a total team score of 189 points), the United States team scored 18 points (with a total team score of 184 points), the Singapore team scored 27 points (with a total team score of 179 points), the Russia team scored 10 points (with a total team score of 161 points), and the Thailand team scored 11 points (with a total team score of 160 points).

The gold medal cutoff for this IMO was set at 28 points (with 54 contestants earning gold medals), the silver medal cutoff was 22 points (with 90 contestants earning silver medals), and the bronze medal cutoff was 16 points (with 137 contestants earning bronze medals).

In this IMO, only one contestant achieved a perfect score of 42 points, namely Lisa Sauermann from Germany.

Problem 5.13 (IMO 54-2, proposed by Australia). A configuration of 4027 points in a plane is called *Colombian* if it consists of 2013 red points and 2014 blue points, and no three of the points of the configuration are collinear. By drawing some lines, the plane is divided into several regions. An arrangement of lines is *good* for a Colombian configuration if the following two conditions are satisfied:

- no line passes through any point of the configuration;
- no region contains points of both colors.

Find the least value of k such that for any Colombian configuration of 4027 points, there is a good arrangement of k lines.

Solution 1. The answer is $k = 2013$.

Firstly, we show that $k \geq 2013$ with an example. We make 2013 red points and 2013 blue points on a circle alternatively. Then there are 4026 arcs on the circle with different endpoint colors. Mark another point in blue on the plane. If k lines are good, then each arc intersects some line and any line intersects the circle at two points at most. Hence there are at least $4026 \div 2 = 2013$ lines.

In the following, we show that there exists a good arrangement of 2013 lines.

First, note that for two points A and B with the same color, we can separate these two points with others by drawing two sufficiently near parallel lines to AB on each side of AB.

Let P be the convex hull of all colored points. Take two adjacent vertices of P, say points A and B. Then other points are located on one side of line AB. If one of them is red, say A. Then we can draw one line to separate A from other points. The remaining 2012 red points can be grouped in 1006 pairs; each pair of red points can be separated by two lines. So, all together 2013 lines meet the requirement.

If A and B are all blue, then they can be separated with other vertices by using a line. The remaining 2012 blue points can be grouped in 1006 pairs; each pair of blue points can be separated by 2 lines. Hence, all together 2013 lines meet the requirements.

Solution 2. We have a more general result as follows.

Suppose there are n points in red or blue in a plane, and no three of them are collinear. Then there exist $\lfloor \frac{n}{2} \rfloor$ lines that meet the requirements.

Proof. We prove it by induction. The conclusion is obvious if $n \leq 2$. For $n \geq 3$, consider a line l passing through points A and B, such that the remaining points are on one side of line l. By the hypotheses of induction, the remaining points have $\lfloor \frac{n}{2} \rfloor - 1$ lines that meet the requirements.

If A and B have the same color or in different colors but are separated by a line, we can separate A and B with other points by a line parallel to l. Thus, $\lfloor \frac{n}{2} \rfloor$ lines meet the requirements. If A and B are in different colors but both in a region R, then the remaining points in the region R do not have a color, say, blue, and we can draw a line to separate the blue point of A or B with other points. Thus, the $\lfloor \frac{n}{2} \rfloor$ lines meet the requirements and it completes the induction.

Remark. We can ask a general problem that substitutes any positive integers m and n with $m \leq n$ for 2013 and 2014, respectively. We denote the solution to the general problem by $f(m, n)$.

We can obtain that $m \leq f(m, n) \leq m + 1$ by the idea of Solution 1. And if m is even, then $f(m, n) = m$. On the other hand, for the case that m is odd, there is an N such that $f(m, n) = m$ for any $m \leq n \leq N$ and $f(m, n) = m + 1$ for any $n > N$.

【Score Situation】 This particular problem saw the following distribution of scores among contestants: 118 contestants scored 7 points, 16 contestants scored 6 points, 12 contestants scored 5 points, 22 contestants scored 4 points, 33 contestants scored 3 points, 65 contestants scored 2 points, 32 contestants scored 1 point, and 229 contestants scored 0 point. The average score of this problem is 2.526, indicating that it had a certain level of difficulty.

Among the top five teams in the team scores, the scores of this problem are as follows: the China team scored 38 points (with a total team score of 208 points), the South Korea team scored 38 points (with a total team score of 204 points), the United States team scored 35 points (with a total team score of 190 points), the Russia team scored 34 points (with a total team score of 187 points), and the North Korea team scored 27 points (with a total team score of 184 points).

The gold medal cutoff for this IMO was set at 31 points (with 45 contestants earning gold medals), the silver medal cutoff was 24 points (with 92 contestants earning silver medals), and the bronze medal cutoff was 15 points (with 141 contestants earning bronze medals).

In this IMO, no contestant achieved a perfect score of 42 points.

Problem 5.14 (IMO 55-6, proposed by Austria). A set of lines in a plane is in *general position* if no two are parallel and no three pass through the same point. A set of lines in general position cut the plane into regions, some of which have finite areas; we call these its *finite regions*.

Prove that for all sufficiently large n, in any set of n lines in general position it is possible to color at least $\sqrt{n}$ of the lines blue in such a way that none of its finite regions has a completely blue boundary.

Note. Results with $\sqrt{n}$ replaced by $c\sqrt{n}$ will be awarded points depending on the value of the constant c.

Proof 1. (the case $c = \sqrt{\frac{1}{2}}$) Denote this set of lines by L. Let B be a maximal subset of L such that no finite region has completely blue boundary when all lines in B are colored blue. Write $|B| = k$, and we shall prove $k \geq \lceil \sqrt{\frac{n}{2}} \rceil$.

We first color all lines in $L \backslash B$ red. Call an intersection point of two blue lines a blue vertex. Hence we have a total of $\binom{k}{2}$ blue points.

Consider a red line l. By the maximality of B, there is a finite region A whose unique red side is on l. Since A has at least three sides, it has at least one blue vertex. Choose a blue vertex of A and associate it with l.

Since every blue vertex belongs to at most 4 finite regions, it is associated to at most four red lines. Thus the total number of red lines does not exceed $4\binom{k}{2}$. Hence the inequality $n - k \leq 2k(k-1)$ holds, that is, $n \leq 2k^2 - k \leq 2k^2$, from which we obtain $k \geq \lceil \sqrt{\frac{n}{2}} \rceil$. The proof is completed.

Note. This is the original problem in the shortlist proposed by Austria. The problem selection committee thought it as a combinatorial problem with medium difficulty. (It is listed as the 5th problem in a total of 9 combinatorial problems.) In addition, the original problem shows that the assertion is true for all $n \geq 3$.

Proof 2. (the case $c = \sqrt{\frac{2}{3}}$) We change the association of red lines and blue vertices a little bit. For a red line l, there is a finite region A whose unique red side is on l. If A is a k-gon, then A has $k - 2$ blue vertices on its boundary. Associate these blue points to l and assign each of these blue vertices the weight $\frac{1}{k-2}$. Thus, the total weight of all blue vertices is $n - k$.

For a fixed blue vertex v, it is easy to see that it belongs to at most four finite regions, among which at most two are triangles. Hence, the total weight assigned to v is not larger than $1 + 1 + \frac{1}{2} + \frac{1}{2} = 3$.

Therefore, $n - k \leq 3\binom{k}{2}$, i.e., $2n \leq 3k^2 - k < 3k^2$. It follows that $k \geq \lceil \sqrt{\frac{2n}{3}} \rceil$.

Proof 3. (the case $c = 1$) We say a point is red if it is the intersection point of a red line and a blue line. For a red line l, there is a finite region A whose unique red side is on l. Denote the vertices of A clockwise by $(p_1, p_2, \ldots, p_s)$ with point p_1 in red and p_2 in blue. Then we say that the red line l corresponds to the red point p_1 and blue point p_2. Now we are to show that any blue point can correspond to at most two red lines. (If this is true, then $n - k \leq 2\binom{k}{2} \Leftrightarrow n \leq k^2$.)

We prove it by contradiction. If not, suppose that there are three red lines l_1, l_2, and l_3 that correspond to a blue point b, and the corresponding red points on red lines l_1, l_2, and l_3 are r_1, r_2, and r_3, respectively. Let b be the intersection point of two blue lines.

Without loss of generality, we may suppose that sides (r_2, b) and (r_3, b) are on one of the two blue lines, and (r_1, b) is the side of region A on the other blue line. Since A has only one red side, A must be a triangle $\triangle r_1 b r_2$. But then red lines l_1 and l_2 pass r_2, and a blue line also passes r_2, which is a contradiction.

【Score Situation】 This particular problem saw the following distribution of scores among contestants: 15 contestants scored 7 points, 1 contestant scored 6 points, 5 contestants scored 5 points, no contestant scored 4 points, 11 contestants scored 3 points, 7 contestants scored 2 points, 7 contestants scored 1 point, and 514 contestants scored 0 point. The average score of this problem is 0.339, indicating that it was extremely difficult.

Among the top five teams in the team scores, the scores of this problem are as follows: the China team scored 24 points (with a total team score of 201 points), the United States team scored 15 points (with a total team score of 193 points), the Chinese Taiwan team scored 13 points (with a total team score of 192 points), the Russia team scored 14 points (with a total team score of 191 points), and the Japan team scored 20 points (with a total team score of 177 points).

The gold medal cutoff for this IMO was set at 29 points (with 49 contestants earning gold medals), the silver medal cutoff was 22 points (with 113 contestants earning silver medals), and the bronze medal cutoff was 16 points (with 133 contestants earning bronze medals).

In this IMO, only three contestants achieved a perfect score of 42 points, namely Jiyang Gao from China, Po-Sheng Wu from Chinese Taiwan, and Alexander Gunning from Australia.

Problem 5.15 (IMO 56-1, proposed by the Netherlands). We say that a finite set S of points in a plane is balanced if, for any two different points A and B in S, there is a point C in S such that $AC = BC$. We say that S is *center-free* if for any three different points A, B, and C in S, there is no point P in S such that $PA = PB = PC$.

(a) Show that for all integers $n \geq 3$, there exists a balanced set consisting of n points.

(b) Determine all integers $n \geq 3$ for which there exists a balanced center-free set consisting of n points.

Proof. (a) For an odd number $n \geq 3$, let S be the set of n vertices of a regular n-gon. We show that S is balanced. Indeed, the points of S are distributed evenly on a circle ω. For any two points $A, B \in S$, they divide ω into two arcs, one of which has an odd number of points in S; the midpoint C of this arc is in S, and $AC = BC$.

For an even number $n \geq 4$, consider the following: Let ω be a circle with center O. Let $k = \frac{n}{2} - 1$ and pick $A_1, \ldots, A_k$ be k points on ω that are very close to each other. To be specific, these k points lie on an arc with central angle $30°$. Rotate each of these points around O exactly $60°$

degrees clockwise and we get k points $B_1, \ldots, B_k$. And rotate A_1 around O counter-clockwise and we get A'. Let

$$S = \{O, A_1, \ldots, A_k, B_1, \ldots, B_k, A'\}.$$

The set S has n distinct points. And we prove that S is balanced. Let A and B be any two distinct points in S. When both A and B are on ω, then $OA = OB$; otherwise, one of A and B is O. By our construction, there is another point $C \in S$ such that $\triangle ABC$ is equilateral, and $CA = CB$.

(b) The answer is all odd numbers $n \geq 3$.

When $n \geq 3$ is an odd number, let S be the set of n vertices of a regular n-gon. We proved in (a) that S is balanced. The circumcenter of any three points in S is the center of the polygon, which is not in S; so S is centre-free.

For any even number $n \geq 4$, we prove that there is no center-free set with n points. Suppose S is a balanced set with n points. For any subset $\{A, B\} \subseteq S$, there are points in S that are equidistant from A and B, we pick any of such points and call it the connection point of $\{A, B\}$. There are C_n^2 binary subsets of S, each determining a connection point.

By the pigeonhole principle, there is a point $P \in S$ which is the connection point for at least $\frac{1}{n}C_n^2 = \frac{1}{2}(n-1)$ binary subsets of S.

Note that n is odd, so P is the connection point for at least $\frac{n}{2}$ binary subsets. The point P is in none of these binary sets, so their elements are from $S \backslash \{P\}$.

Since $\frac{n}{2} \times 2 = n > n - 1$, two of these binary subsets overlap, say, $\{A, B\}$ and $\{A, C\}$. Hence $PA = PB = PC$, and S is not center-free.

【Score Situation】 This particular problem saw the following distribution of scores among contestants: 265 contestants scored 7 points, 20 contestants scored 6 points, 12 contestants scored 5 points, 72 contestants scored 4 points, 21 contestants scored 3 points, 5 contestants scored 2 points, 89 contestants scored 1 point, and 93 contestants scored 0 point. The average score of this problem is 4.307, indicating that it was simple.

Among the top five teams in the team scores, the scores of this problem are as follows: the United States team scored 42 points (with a total team score of 185 points), the China team scored 42 points (with a total team score of 181 points), the South Korea team scored 42 points (with a total team score of 161 points), the North Korea team scored 42 points (with a total team score of 156 points), and the Vietnam team scored 35 points (with a total team score of 151 points).

The gold medal cutoff for this IMO was set at 26 points (with 39 contestants earning gold medals), the silver medal cutoff was 19 points (with 100 contestants earning silver medals), and the bronze medal cutoff was 14 points (with 143 contestants earning bronze medals).

In this IMO, only one contestant achieved a perfect score of 42 points, namely Zhuo Qun Alex Song from Canada.

Problem 5.16 (IMO 57-6, proposed by Czech Republic). There are $n \geq 2$ line segments in a plane such that every two segments cross, and no three segments meet at a point. Geoff must choose an endpoint of each segment and place a frog on it, facing the other endpoint. Then he will clap his hands $n - 1$ times. Every time he claps, each frog will immediately jump forward to the next intersection point on its segment. Frogs never change the direction of their jumps. Geoff wishes to place the frogs in such a way that no two of them will ever occupy the same intersection point at the same time.

(a) Prove that Geoff can always fulfil his wish if n is odd.
(b) Prove that Geoff can never fulfil his wish if n is even.

Proof. Pick a circle ω that is big enough to contain all the segments in its interior. Extend each segment on both sides until it meets ω at two points. We may assume that the game is played on these new and extended segments.

So, we have n segments as n chords of ω, any two of them intersect inside ω, and no three of them meet at a point. We label the $2n$ endpoints of the n chords as $A_1, A_2, A_3, \ldots, A_{2n}$ according to their clockwise order on ω.

(a) Geoff can place the frogs on points $A_1, A_3, \ldots, A_{2n-1}$. First of all, for any chord there are $n - 1$ points on each side of it, so these chords are $A_i A_{i+n}$ with $i = 1, 2, \ldots, n$. Thus, indeed Geoff placed exactly one frog on each chord. In order to prove that no two frogs will occupy the same intersection, we observe any two frogs, initially placed on A_i and A_{i+2k} with $1 \leq k < \frac{n}{2}$, all the indices modulo $2n$.

Let P be the intersection of $A_i A_{i+n}$ and $A_{i+2k} A_{i+2k+n}$. It is enough to prove that the segments $A_i P$ and $A_{i+2k} P$ contain different numbers of intersection points. Each of the chords $A_j A_{j+n}$, where $j = i + 1, i + 2, \ldots,$ $i+2k-1$ intersects exactly one of $A_i P$ and $A_{i+2k} P$; every other chord either intersects both $A_i P$ and $A_{i+2k} P$ or none of them. So, the total number of the intersection points on $A_i P$ and $A_{i+2k} P$ is odd; we do not have the same number on them.

(b) There must be two adjacent indices A_i and A_{i+1} picked as initial positions by Geoff; otherwise, among the $2n$ positions, Geoff must have picked every other position, and since n is even, he must have picked some A_i and A_{i+n}, which are on the same chord.

Let P be the intersection of $A_i A_{i+n}$ and $A_{i+1} A_{i+n+1}$. Each of the other chords either intersects both $A_i P$ and $A_{i+1} P$ or none of them. Hence, we have the same number of intersection points on $A_i P$ and $A_{i+1} P$; the two frogs initially placed on A_i and A_{i+1} will come to P at the same time.

【Score Situation】 This particular problem saw the following distribution of scores among contestants: 37 contestants scored 7 points, 4 contestants scored 6 points, 4 contestants scored 5 points, 4 contestants scored 4 points, 39 contestants scored 3 points, 9 contestants scored 2 points, 31 contestants scored 1 point, and 474 contestants scored 0 points. The average score of this problem is 0.806, indicating that it was extremely difficult.

Among the top five teams in the team scores, the scores of this problem are as follows: the United States team scored 25 points (with a total team score of 214 points), the South Korea team scored 26 points (with a total team score of 207 points), the China team scored 28 points (with a total team score of 204 points), the Singapore team scored 20 points (with a total team score of 196 points), and the Chinese Taiwan team scored 21 points (with a total team score of 175 points).

The gold medal cutoff for this IMO was set at 29 points (with 44 contestants earning gold medals), the silver medal cutoff was 22 points (with 101 contestants earning silver medals), and the bronze medal cutoff was 16 points (with 135 contestants earning bronze medals).

In this IMO, a total of six contestants achieved a perfect score of 42 points.

Problem 5.17 (IMO 61-6, proposed by Chinese Taiwan). Prove that there exists a positive constant c such that the following statement is true: Consider an integer $n > 1$ and a set S of n points in a plane such that the distance between any two different points in S is at least 1. Then there is a line l separating S such that the distance from any point of S to l is at least $cn^{-\frac{1}{3}}$.

(A line l separates a set of points S if some segment joining two points in S crosses l.)

Note. Weaker results with $cn^{-\frac{1}{3}}$ replaced by $cn^{-\alpha}$ may be awarded points depending on the value of the constant $\alpha > \frac{1}{3}$.

Proof. We prove that the statement is true for $c = 0.1$.

Let $\delta = cn^{-\frac{1}{3}}$, and for a set of points S in the plane and line l, we use $\delta(S, l)$ to denote the minimal distance from a point in S to l. Assume that the assertion is not true. Then there exists a set S of n points in the plane ($n \geq 2$), such that any line l separating S satisfies $\delta(S, l) < \delta$. We choose two points A and B in S such that the distance between them is the maximal. Let $d = |AB|$. Then apparently $d \geq 1$.

Now we construct a coordinate system such that A is the origin and $\overrightarrow{AB}$ coincides with the positive direction of the x-axis. Suppose the x-coordinates of points in S are $d_1 \leq d_2 \leq \cdots \leq d_n$. Since the points in S are inside the intersection of the following two disks:

$$D_A = \{P \in \mathbf{R}^2| \ |PA| \leq d\} \quad \text{and} \quad D_B = \{P \in \mathbf{R}^2| \ |PB| \leq d\},$$

it follows that all the x-coordinates lie in the interval $[0, d]$, so $d_1 = 0$ and $d_n = d$.

If there exists $1 \leq i \leq n - 1$ such that $d_{i+1} - d_i \geq 2\delta$, then line $l : x = \frac{d_i + d_{i+1}}{2}$ separates S and $\delta(S, l) \geq \delta$, which is contradictory to the assumption.

Hence, $d_{i+1} - d_i < 2\delta$ for every $1 \leq i \leq n - 1$. Now consider the points of S in the strip $0 \leq x \leq \frac{1}{2}$ and let them be $P_1, P_2, \ldots, P_k$, where P_i has coordinates (d_i, y_i), $i = 1, 2, \ldots, k$. Since $d_i \leq 2(i - 1)\delta$, at least $\lceil \frac{1}{4\delta} \rceil$ numbers in $d_1, d_2, \ldots, d_n$ belong to interval $[0, \frac{1}{2}]$, i.e., $k \geq \lceil \frac{1}{4\delta} \rceil \geq \frac{1}{4\delta}$.

For $1 \leq i < j \leq k$, since $|d_j - d_i| \leq \frac{1}{2}$ and $|P_i P_j| \geq 1$, it follows that $|y_j - y_i| \geq \frac{\sqrt{3}}{2}$. Note that this implies that the difference between any two of $y_1, y_2, \ldots, y_k$ is at least $\frac{\sqrt{3}}{2}$, so

$$\max_{1 \leq i \leq k} y_i - \min_{1 \leq i \leq k} y_i \geq (k - 1)\frac{\sqrt{3}}{2}.$$

The intersection of the strip $0 \leq x \leq \frac{1}{2}$ and D_B is a bow-shaped region, whose highest point and lowest point are the intersection points of $x = \frac{1}{2}$ and the circle $(x - d)^2 + y^2 = d^2$, which are $(\frac{1}{2}, \pm\sqrt{d - \frac{1}{4}})$. Consequently,

$$\max_{1 \leq i \leq k} y_i - \min_{1 \leq i \leq k} y_i \leq 2\sqrt{d - \frac{1}{4}} < 2\sqrt{d}.$$

In combination, together with $k \geq \frac{1}{4\delta} > 2$, we obtain

$$2\sqrt{d} > (k - 1)\frac{\sqrt{3}}{2} \geq \frac{\sqrt{3}}{4}k \geq \frac{\sqrt{3}}{16\delta}.$$

Squaring both sides, and using $d < 2n\delta$, we have

$$8n\delta > 4d > \frac{3}{256\delta^2}.$$

Equivalently, $2048n\delta^3 > 3$. Since $\delta = cn^{-\frac{1}{3}}$, we have $2048n\delta^3 = 2048c^3 > 3$, which is not true for $c = 0.1$. Therefore, the converse assumption cannot hold, which proves our conclusion.

【Score Situation】This particular problem saw the following distribution of scores among contestants: 4 contestants scored 7 points, 1 contestant scored 6 points, 1 contestant scored 5 points, 1 contestant scored 4 points, 1 contestant scored 3 points, 1 contestant scored 2 points, 126 contestants scored 1 point, and 481 contestants scored 0 point. The average score of this problem is 0.282, indicating that it was extremely difficult.

Among the top five teams in the team scores, the scores of this problem are as follows: the China team scored 31 points (with a total team score of 215 points), the Russia team scored 28 points (with a total team score of 185 points), the United States team scored 15 points (with a total team score of 183 points), the South Korea team scored 22 points (with a total team score of 175 points), and the Thailand team scored 15 points (with a total team score of 174 points).

The gold medal cutoff for this IMO was set at 31 points (with 49 contestants earning gold medals), the silver medal cutoff was 24 points (with 112 contestants earning silver medals), and the bronze medal cutoff was 15 points (with 155 contestants earning bronze medals).

In this IMO, only one contestant achieved a perfect score of 42 points, namely Jinmin Li from China.

5.2.2   Shape problems

Problem 5.18 (IMO 12-6, proposed by the Soviet Union). In a plane there are 100 points, no three of which are collinear. Consider all possible triangles having these points as vertices. Prove that no more than 70% of these triangles are acute-angled.

Proof. We first show that for any 5 points in the plane with no three collinear, there exist three non-acute triangles determined by these points.

If the convex hull of P_1, P_2, P_3, P_4, P_5 is a triangle, say $\triangle P_1 P_2 P_3$, then P_4 and P_5 are in the interior of $\triangle P_1 P_2 P_3$. At least two of $\angle P_1 P_4 P_2$, $\angle P_2 P_4 P_3$, and $\angle P_3 P_4 P_1$ are not acute. Hence at least two of $\triangle P_1 P_4 P_2$, $\triangle P_2 P_4 P_3$, and $\triangle P_3 P_4 P_1$ are non-acute triangles. Similarly, at least two of $\triangle P_1 P_5 P_2$, $\triangle P_2 P_5 P_3$, and $\triangle P_3 P_5 P_1$ are non-acute triangles. In this case, we have at least four non-acute triangles.

If the convex hall of P_1, P_2, P_3, P_4, P_5 is a quadrilateral, say $P_1 P_2 P_3 P_4$, then P_5 is in the interior of $P_1 P_2 P_3 P_4$, as shown in Figure 5.7. Since at least one of $\angle P_1 P_2 P_3$, $\angle P_2 P_3 P_4$, $\angle P_3 P_4 P_1$, and $\angle P_4 P_1 P_2$ is not acute, we have at least one non-acute triangle in $\triangle P_1 P_2 P_3$, $\triangle P_2 P_3 P_4$, $\triangle P_3 P_4 P_1$, and $\triangle P_4 P_1 P_2$.

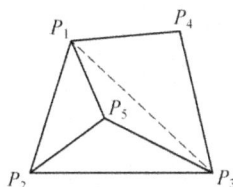

Figure 5.7

Assume without loss of generality that P_5 is in the interior of $\triangle P_1P_2P_3$. Then at least two of $\triangle P_1P_2P_5$, $\triangle P_2P_3P_5$, $\triangle P_3P_1P_5$ are non-acute. In this case, we have at least three non-acute triangles.

If the convex hull is a pentagon $P_1P_2P_3P_4P_5$, then at least two of $\angle P_1P_2P_3$, $\angle P_2P_3P_4$, $\angle P_3P_4P_5$, $\angle P_4P_5P_1$, and $\angle P_5P_1P_2$ are non-acute since these five angles add up to $540°$, as shown in Figure 5.8. If exactly two of them are non-acute, then there exist two adjacent vertices with acute angles, say $\angle P_5P_1P_2$ and $\angle P_1P_2P_3$. Hence at least one of $\angle P_1P_5P_3$ and $\angle P_2P_3P_5$ is non-acute. In this case, again we have at least three non-acute triangles.

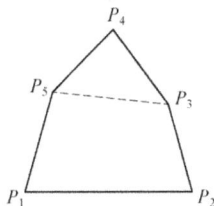

Figure 5.8

Thus, in any configuration of five points, at least three triangles are non-acute.

Since every triangle belongs to C_{97}^2 quintuples, there are at least $\dfrac{3C_{100}^5}{C_{97}^2}$ non-acute triangles. Therefore, the number of acute triangles is at most

$$C_{100}^3 - \frac{3C_{100}^5}{C_{97}^2}.$$

Since

$$\frac{C_{100}^3 - \frac{3C_{100}^5}{C_{97}^2}}{C_{100}^3} = 1 - \frac{3C_{100}^5}{C_{100}^3 C_{97}^2} = \frac{7}{10} = 70\%,$$

the assertion is proved.

【Score Situation】This particular problem saw the following distribution of scores among contestants: 7 contestants scored 8 points, 2 contestants scored 7 points, no contestant scored 6 points, 1 contestant scored 5 points, no contestant scored 4 points, no contestant scored 3 points, 3 contestants scored 2 points, 3 contestants scored 1 point, and 19 contestants scored 0 point. The average score of this problem is 2.400, indicating that it had a certain level of difficulty.

Among the top five teams in the team scores, the Hungary team achieved a total score of 233 points, the German Democratic Republic team achieved a total score of 221 points, the Soviet Union team achieved a total score of 221 points, the Yugoslavia team achieved a total score of 209 points, and the Romania team achieved a total score of 208 points.

The gold medal cutoff for this IMO was set at 37 points (with 7 contestants earning gold medals), the silver medal cutoff was 30 points (with 11 contestants earning silver medals), and the bronze medal cutoff was 19 points (with 40 contestants earning bronze medals).

In this IMO, only three contestants achieved a perfect score of 40 points, namely Wolfgang Burmeister from the German Democratic Republic, Imre Ruzsa from Hungary, and Andrei Hodulev from the Soviet Union.

Problem 5.19 (IMO 31-6, proposed by the Netherlands). Prove that there exists a convex 1990-gon with the following two properties:

(a) All angles are equal.
(b) The lengths of the 1990 sides are the numbers $1^2, 2^2, \ldots, 1989^2, 1990^2$ in some order.

Proof. The problem is equivalent to the existence of a permutation of $1^2, 2^2, \ldots, 1990^2$, denoted by $a_1, a_2, \ldots, a_{1990}$, satisfying

$$\sum_{k=1}^{1990} a_k (\cos k\theta + \mathrm{i} \sin k\theta) = 0, \tag{1}$$

where $\theta = \frac{2\pi}{1990} = \frac{\pi}{995}$.

We take $\{(a_{2k-1}, a_{2k-1+995}) \mid k = 1, 2, \ldots, 995\} = \{((2n-1)^2, (2n)^2) \mid n = 1, 2, \ldots, 995\}$, where by convention, we assume $a_{j+1990} = a_j$ for $j = 1, 2, \ldots$. Then identity (1) becomes

$$\sum_{k=1}^{995} b_k (\cos 2k\theta + \mathrm{i} \sin 2k\theta) = 0, \tag{2}$$

where $b_1, b_2, \ldots, b_{995}$ is a permutation of $(2n)^2 - (2n-1)^2 = 4n - 1(n = 1, 2, \ldots, 995)$. Set

$$S_r = \sum_{t=0}^{4} b_{199t+5r}(\cos 2(199t + 5r)\theta + i \sin 2(199t + 5r)\theta)$$

$$(r = 1, 2, \ldots, 199, \ t = 0, 1, 2, 3, 4),$$

where $b_{199t+5r} = 4(5r + t) - 17$ (assuming $b_{j+995} = b_j$).

We note that $b_{199t+5r}(r = 1, 2, \ldots, 199, t = 0, 1, 2, 3, 4)$ form a complete residue system modulo 995 and $4(5r + t) - 17(r = 1, 2, \ldots, 199, t = 0, 1, 2, 3, 4)$ take each value $(2n)^2 - (2n-1)^2 = 4n - 1 \quad (n = 1, 2, \ldots, 995)$ exactly once.

Thus, $b_1, b_2, \ldots, b_{995}$ is a permutation of $(2n)^2 - (2n-1)^2 = 4n - 1(n = 1, 2, \ldots, 995)$.

Using $\sum_{t=0}^{4} (\cos(2 \times 199t\theta) + i \sin(2 \times 199t\theta)) = 0$, we obtain that, for $1 \leq r \leq 199$,

$$S_r = 4(\cos 10r\theta + i \sin 10r\theta) \sum_{t=0}^{4} t(\cos(2 \times 199t\theta) + i \sin(2 \times 199t\theta))$$

$$= 4(\cos 10r\theta + i \sin 10r\theta)S,$$

where

$$S = \sum_{t=0}^{4} t(\cos(2 \times 199t\theta) + i \sin(2 \times 199t\theta)).$$

Thus

$$\sum_{k=1}^{995} b_k(\cos 2k\theta + i \sin 2k\theta) = 4S \sum_{r=1}^{199} (\cos 10r\theta + i \sin 10r\theta) = 0,$$

that is, identity (2) holds.

【Score Situation】 This particular problem saw the following distribution of scores among contestants: 17 contestants scored 7 points, 3 contestants scored 6 points, 2 contestants scored 5 points, 18 contestants scored 4 points, 31 contestants scored 3 points, 29 contestants scored 2 points, 93 contestants scored 1 point, and 115 contestants scored 0 point. The average score of this problem is 1.503, indicating that it was relatively challenging.

Among the top five teams in the team scores, the scores of this problem are as follows: the China team scored 29 points (with a total team score of 230 points), the Soviet Union team scored 28 points (with a total team score of 193 points), the United States team scored 17 points (with a total team score of 174 points), the Romania team scored 11 points (with a total team score of 171 points), and the France team scored 20 points (with a total team score of 168 points).

The gold medal cutoff for this IMO was set at 34 points (with 23 contestants earning gold medals), the silver medal cutoff was 23 points (with 56 contestants earning silver medals), and the bronze medal cutoff was 16 points (with 76 contestants earning bronze medals).

In this IMO, a total of four contestants achieved a perfect score of 42 points.

Problem 5.20 (IMO 47-2, proposed by Serbia-Montenegro). Let P be a regular 2006-gon. A diagonal of P is called *good* if its endpoints divide the boundary of P into two parts, each composed of an odd number of sides of P. The sides of P are also called *good*.

Suppose P has been dissected into triangles by 2003 diagonals, no two of which have a common point in the interior of P. Find the maximum number of isosceles triangles having two good sides that could appear in such a configuration.

Solution. Let $\triangle ABC$ be a triangle in the partition. Here arc AB denotes the part of the perimeter of P outside the triangle and between points A and B, and similarly for arc BC and arc CA. Let a, b, and c be the number of sides on arc AB, arc BC, and arc CA, respectively. Note that $a + b + c = 2006$. By a parity check, if an isosceles triangle has two good sides, then these sides must be two equal sides.

We call such isosceles triangles *good*.

Let one of the *good* triangles be $\triangle ABC$ with $AB = AC$. Also, inscribe our polygon into a circle. If there is another *good* triangle in arc AB, then the two equal good sides cut off an even number of sides in arc AB. Since there is an odd number of sides in arc AB, there must be one side not belonging to any good triangle. The same holds for arc AC.

So, every good triangle corresponds to at least two sides of P. Hence there are no more than $2006 \div 2 = 1003$ *good* triangles.

And this bound can be achieved. Let $P = A_1 A_2 \cdots A_{2006}$. Draw diagonals between $A_1 A_{2k+1} (1 \le k \le 1002)$ and $A_{2k+1} A_{2k+3} (1 \le k \le 1001)$. This gives us the required 1003 *good* triangles.

【Score Situation】This particular problem saw the following distribution of scores among contestants: 78 contestants scored 7 points, 8 contestants scored 6 points, 8 contestants scored 5 points, 15 contestants scored 4 points, 3 contestants scored 3 points, no contestant scored 2 points, 210 contestants scored 1 point, and 176 contestants scored 0 point. The average score of this problem is 1.833, indicating that it was relatively challenging.

Among the top five teams in the team scores, the scores of this problem are as follows: the China team scored 42 points (with a total team score of 214 points), the Russia team scored 36 points (with a total team score of 174 points), the South Korea team scored 20 points (with a total team score of 170 points), the Germany team scored 24 points (with a total team score of 157 points), and the United States team scored 41 points (with a total team score of 154 points).

The gold medal cutoff for this IMO was set at 28 points (with 42 contestants earning gold medals), the silver medal cutoff was 19 points (with 89 contestants earning silver medals), and the bronze medal cutoff was 15 points (with 122 contestants earning bronze medals).

In this IMO, only three contestants achieved a perfect score of 42 points, namely Zhiyu Liu from China, Iurie Boreico from Moldova, and Alexander Magazinov from Russia.

5.2.3 Covering, embedding, partitioning, and patching problems

Problem 5.21 (IMO 16-4, proposed by Bulgaria). Consider decompositions of an 8×8 chessboard into p non-overlapping rectangles subject to the following conditions:

(i) Each rectangle has as many white squares as black squares.
(ii) If a_i is the number of white squares in the i-th rectangle, then
$$a_1 < a_2 < \cdots < a_p.$$

Find the maximum value of p for which such a decomposition is possible. For this value of p, determine all possible sequences $a_1, a_2, \ldots, a_p$.

Solution. As shown in Figure 5.9, the board has 32 white squares, and we have

$$a_1 + a_2 + \cdots + a_p = 32.$$

Figure 5.9

Since $a_1 \geq 1, a_2 \geq 2, \ldots, a_p \geq p$,

$$32 \geq 1 + 2 + \cdots + p = \frac{p(p+1)}{2}.$$

It follows that $p^2 + p \leq 64$, so $p \leq 7$. This shows that there can be at most seven rectangles in a decomposition of the type required.

To show that the decomposition into seven rectangles exists and to find them all, we seek seven distinct positive integers whose sum is 32. Here is a complete list:

  (i) $1 + 2 + 3 + 4 + 5 + 6 + 11$;
 (ii) $1 + 2 + 3 + 4 + 5 + 7 + 10$;
(iii) $1 + 2 + 3 + 4 + 5 + 8 + 9$;
(iv) $1 + 2 + 3 + 4 + 6 + 7 + 9$;
 (v) $1 + 2 + 3 + 5 + 6 + 7 + 8$.

Now case (i) is impossible because no rectangle with 22 squares can be cut from an 8×8 board. All other cases are possible and the corresponding subdivisions are shown in Figure 5.10.

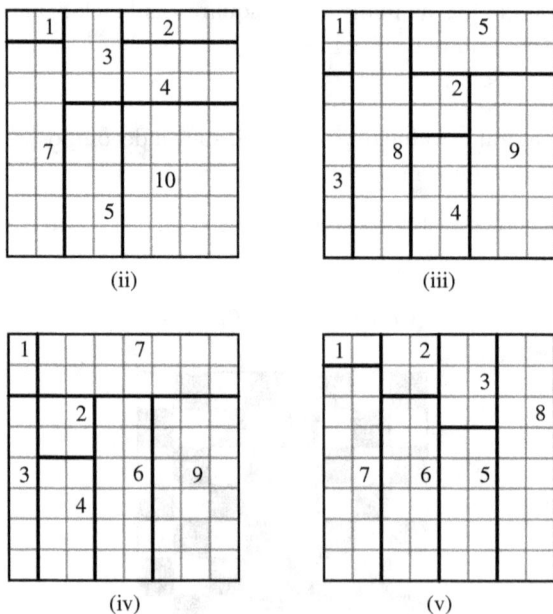

Figure 5.10

scored 1 point, and 4 contestants scored 0 point. The average score of this problem is 5.064, indicating that it was simple.

Among the top five teams in the team scores, the scores of this problem are as follows: the Soviet Union team scored 46 points (with a total team score of 256 points), the United States team scored 43 points (with a total team score of 243 points), the Hungary team scored 44 points (with a total team score of 237 points), the German Democratic Republic team scored 45 points (with a total team score of 236 points), and the Yugoslavia team scored 47 points (with a total team score of 216 points).

The gold medal cutoff for this IMO was set at 38 points (with 10 contestants earning gold medals), the silver medal cutoff was 30 points (with 24 contestants earning silver medals), and the bronze medal cutoff was 23 points (with 37 contestants earning bronze medals).

In this IMO, a total of six contestants achieved a perfect score of 40 points.

Problem 5.22 (IMO 45-3, proposed by Estonia). Define a "hook" to be a figure made up of six unit squares as shown in Figure 5.11, or any of the figures obtained by applying rotations and reflections to this figure.

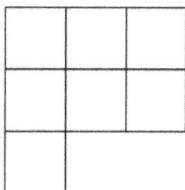

Figure 5.11

Determine all $m \times n$ rectangles that can be covered without gaps and overlaps with hooks such that

- the rectangle is covered without gaps and overlaps;
- no part of a hook covers an area outside the rectangle.

Solution 1. The positive integers m and n should satisfy one of the following conditions: (1) $3 \mid m$ and $4 \mid n$ (or vice versa); (2) one of m and n is divisible by 12 and one is not less than 7.

A figure is obtained by applying rotations and reflections to another figure. We regard the two figures as equivalent.

Label the six unit squares of the hook as shown in Figure 5.12. The shaded square must belong to another hook, and it is adjacent to only one square of this other hook. Then the only possibility of the shaded square is 1 or 6.

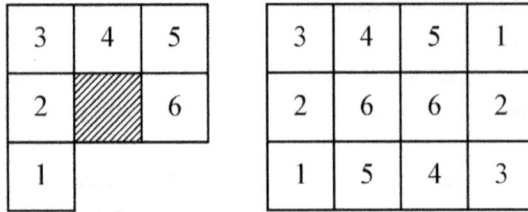

Figure 5.12

(i) If it is 6, then two hooks form a 3×4 rectangle. We call it (1).

(ii) If it is 1, then there are two cases.

It is easy to see that the shaded square cannot be covered in the first diagram as shown in Figure 5.13. Hence the latter is true. We call it (2).

Thus, in a tessellation, all hooks are matched into pairs. Each pair forms (1) or (2).

There are 12 squares in (1) and (2). Hence $12 \mid mn$. Now we consider three cases separately.

(1) $3 \mid m$ and $4 \mid n$ (or vice versa).

Without loss of generality, we may assume $m = 3m_0$ and $n = 4n_0$. Then $m_0 n_0$ rectangles of type (1) form an $m_0 \times n_0$ rectangle. Since two hooks cover a 3×4 rectangle, an $m \times n$ rectangle can be covered with hooks.

(2) $12 \mid m$ or $12 \mid n$.

Without loss of generality, we may assume $12 \mid m$. If $3 \mid n$ or $4 \mid n$, then the question reduces to (1).

Assume that n is not divisible by 3 nor by 4. If a tessellation exists, then there is at least one of (1) and (2) in it, so $n \geq 3$. Hence $n \geq 5$ because $3 \nmid n$ and $4 \nmid n$. Since the square at the corners can belong to either (1) or (2), it follows from $n \geq 5$ that the squares at the adjacent corners cannot belong to the same type (1) or (2). Hence $n \geq 6$. Since n is not divisible by 3 and 4, we have $n \geq 7$.

We now show that if $n \geq 7$ and n is not divisible by 3 and 4, a tessellation exists.

If $n \equiv 1 \pmod 3$, then $n = 4 + 3t$ ($t \in \mathbf{N}^*$). Together with (1), we see that if $12 \mid m$, then an $m \times 3t$ rectangle and an $m \times 4$ rectangle can be covered with hooks. So, the problem can be solved.

If $n \equiv 2 \pmod 3$, then $n = 8 + 3t$ ($t \in \mathbf{N}^*$). Together with (1), we know that if $12 \mid m$, then each of the $m \times 8$ and $m \times 3t$ rectangles can be covered with hooks. The problem is solved.

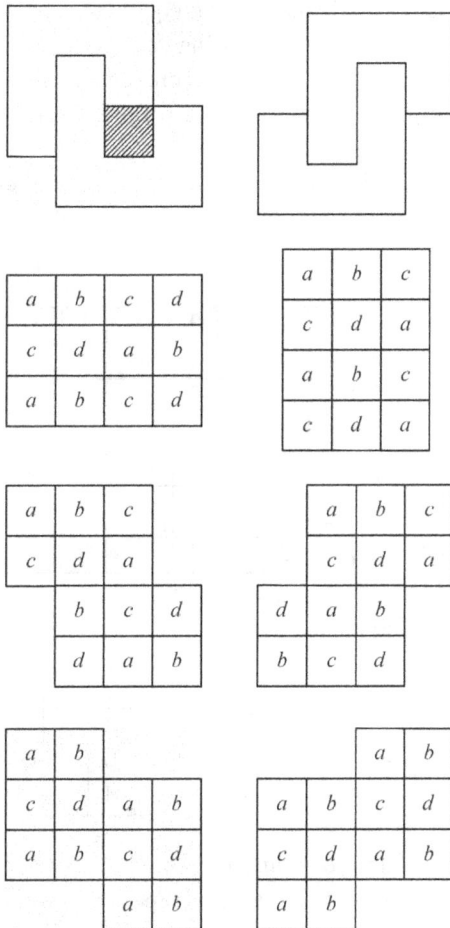

Figure 5.13

(3) $12 \mid mn$, but neither m nor n is divisible by 4. Now $2 \mid m$ and $2 \mid n$. We may assume without loss of generality that $m = 6m_0$, $n = 2n_0$, and neither m_0 nor n_0 is divisible by 2. We will prove that if these conditions are satisfied, then an $m \times n$ rectangle cannot be covered with hooks.

Consider coloring the columns of an $m \times n$ matrix with black and white colors alternately. Then the number of the black squares equals that of the white ones. One (2) always covers six black squares. A horizontal (1) always covers six black squares. A vertical (1) covers either eight black squares and four white ones, or four black squares and eight white ones.

Since the number of the black squares equals that of the white ones, the number of (1) is the same in the preceding two cases. Hence the total number of a vertical (1) is even. Using the same argument as above (coloring the rows alternately), we obtain that the total number of a horizontal (1) is even.

Consider classifying the squares of the $m \times n$ rectangle into four types marked 1, 2, 3, and 4 as shown below. The number of squares of each type is equal to $\frac{mn}{4}$.

1	2	3	4	1	2	...	1	2
3	4	1	2	3	4	...	3	4
1	2	3	4	1	2	...	1	2
						...		
						...		
⋮	⋮	⋮	⋮	⋮	⋮		⋮	⋮
3	4	1	2	3	4		3	4

From the two diagrams below,

a	b	c	d
c	d	a	b
a	b	c	d

a	b	c
c	d	a
a	b	c
c	d	a

we obtain that the number of a and c covered by (1) is the same, so is for b and d. Hence the number of squares of type 1 covered by (1) equals that of type 3.

a	b	c	
c	d	a	
	b	c	d
	d	a	b

(i)

	a	b	c
	c	d	a
d	a	b	
b	c	d	

(ii)

a	b		
c	d	a	b
a	b	c	d
		a	b

(iii)

		a	b
a	b	c	d
c	d	a	b
a	b		

(iv)

The number of squares of type 1 covered by (i) or (ii) equals that of type 3. The difference between the number of squares of type 1 and type 3 covered by (iii) or (iv) is 2. There are two cases: the number of squares of type 1 is 2 more than that of type 3, or vice versa.

Since the number of squares of type 1 equals that of type 3 in the rectangle, the fray nancy of the two cases is the same. Hence the total number of (iii) and (iv) is even.

Consider classifying the squares of the $m \times n$ rectangle as shown below.

1	3	1	
2	4	2	
3	1	3	
4	2	4	

Similarly, the total number of (i) and (ii) is even. Then the number of (2) is even. So, there is an even number of (1) and (2). Hence $24 \mid m \times n$, contrary to the assumption that neither m nor n is divisible by 4.

Solution 2. The positive integers m and n should satisfy one of the following conditions: (1) $3 \mid m$ and $4 \mid n$ (or vice versa); (2) $12 \mid m$ and $n \neq 1, 2, 5$ (or vice versa).

Consider a covering of an $m \times n$ rectangle satisfying the conditions. For any hook A, there is a unique hook B covering the "inner" square of A with one of its "tailed" squares. In turn, the "inner" square of B must be covered by a "tailed" square of A.

Thus, in a tessellation, all hooks are matched into pairs. There are only two possibilities to place B so that it does not overlap with A and no gap occurs. In one case, A and B form a 3×4 rectangle; in the other, their union is an octagonal shape, with sides of length 3, 2, 1, 2, 3, 2, 1, 2, respectively.

So, an $m \times n$ rectangle can be covered with hooks if and only if it can be covered with the 12-square tiles described above. Suppose that such a tessellation exists; then mn is divisible by 12. We now show that one of m and n is divisible by 4.

Assume on the contrary that this is not the case. Then m and n are both even, because mn is divisible by 4. Imagine that the rectangle is divided into unit squares, with the rows and columns labeled $1, 2, \ldots, m$ and $1, 2, \ldots, n$.

Write 1 in the square (i, j) if exactly one of i and j is divisible by 4, and 2, if i and j are both divisible by 4. Since the number of squares in each row and column is even, the sum of all numbers written is also even.

Now, it is easy to check that a 3×4 rectangle always covers numbers with sums 3 or 7; and the other 12-square shape always covers numbers with sums 5 or 7. Consequently, the total number of 12-square shapes is even.

But then mn is divisible by 24, and hence by 8, contrary to the assumption that m and n are not divisible by 4.

Notice that neither m nor n can be 1, 2, or 5 (any attempt to place tiles along a side of length 1, 2, or 5 fails). We infer that if a tessellation is possible, then one of m and n is divisible by 3, one is divisible by 4, and $m, n \notin \{1, 2, 5\}$.

Conversely, we shall prove that if these conditions are satisfied, then a tessellation is possible (using only 3×4 rectangles). The result is immediate if 3 divides m and 4 divides n (or vice versa). Let m be divisible by 12 and $n \notin \{1, 2, 5\}$ (or vice versa).

Without loss of generality, we may assume that neither 3 nor 4 divides n. Then $n \geq 7$. In addition, between $n - 4$ and $n - 8$, at least one can be divisible by 3. Hence the rectangle can be partitioned into $m \times 3$ and $m \times 4$ rectangles, which are easy to cover, in fact with only 3×4 tiles again.

【Score Situation】This particular problem saw the following distribution of scores among contestants: 11 contestants scored 7 points, no contestant scored 6 points, 1 contestant scored 5 points, 15 contestants scored 4 points, 30 contestants scored 3 points, 80 contestants scored 2 points, 100 contestants scored 1 point, and 249 contestants scored 0 point. The average score of this problem is 1.012, indicating that it was relatively challenging.

Among the top five teams in the team scores, the scores of this problem are as follows: the China team scored 21 points (with a total team score of 220 points), the United States team scored 20 points (with a total team score of 212 points), the Russia team scored 25 points (with a total team score of 205 points), the Vietnam team scored 13 points (with a total team score of 196 points), and the Bulgaria team scored 22 points (with a total team score of 194 points).

The gold medal cutoff for this IMO was set at 32 points (with 45 contestants earning gold medals), the silver medal cutoff was 24 points (with 78 contestants earning silver medals), and the bronze medal cutoff was 16 points (with 120 contestants earning bronze medals).

In this IMO, a total of four contestants achieved a perfect score of 42 points.

5.2.4   *Lattice point and grid problems*

Problem 5.23 (IMO 27-6, proposed by the German Democratic Republic). One is given a finite set of points in a plane, each point having integer coordinates. Is it always possible to color some of the points in the set red and the remaining points white in such a way that for any straight line L parallel to either one of the coordinate axes the difference (in absolute value) between the numbers of white points and red points on L is not greater than 1?

Solution. The answer is yes. We prove this by induction on the number of points in the given set, denoted by A.

When $n = 1$, there is nothing to prove.

Fix $n \geq 2$, and assume that every set of cardinality less than n can be colored in the desired manner. Let A be an n-point set. Consider two cases.

(1) There exists a horizontal or vertical line l with an odd number of points of the set A. Choose one of these points, call it P, and remove it from A. Color the points of the remaining $(n-1)$-element set red and white, according to the inductive hypothesis. An even number of points remain on line l (possibly empty), half of them red and the other half white.

Let l' be the line passing through P and perpendicular to l. The number of red points (from the set $A\backslash\{P\}$) on l' differs from the number of white points by at most 1. So, we can choose a color for the point P without violating this property. On line l we now have one red point more or less than there are white points.

(2) Every line, horizontal or vertical, contains an even number of points of the set A. Choose arbitrarily a vertical line m which intersects A. Remove the points on m and color the remaining points according to the inductive hypothesis. Now for each vertical line other than m, there are an equal number of red and white points. There are two cases for horizontal lines.

If a horizontal line does not contain any point in A on line m, then it has an equal number of red and white points by the inductive hypothesis. If a horizontal line does contain a point in A on line m, then we have one red point more or less on this horizontal line. The number of such horizontal lines with one more red point equals the number of such horizontal lines with one more white point since totally we have an equal number of red points and white points counted vertically.

For each such horizontal line, we color the point, its intersection with m red, if the white points were one more than the red points, and color it white otherwise. Thus, on line m, we also have an equal number of red and white points.

It follows by induction that the desired coloring exists for every finite set.

【Score Situation】 This particular problem saw the following distribution of scores among contestants: 43 contestants scored 7 points, 2 contestants scored 6 points, 5 contestants scored 5 points, 3 contestants scored 4 points, 5 contestants scored 3 points, 7 contestants scored 2 points, 30 contestants scored 1 point, and 115 contestants scored 0 point. The average score of this problem is 1.948, indicating that it was relatively challenging.

Among the top five teams in the team scores, the scores of this problem are as follows: the Soviet Union team scored 36 points (with a total team score of 203 points), the United States team scored 33 points (with a total team score of 203 points), the Germany team scored 30 points (with a total team score of 196 points), the China team scored 32 points (with a total team score of 177 points), and the German Democratic Republic team scored 17 points (with a total team score of 172 points).

The gold medal cutoff for this IMO was set at 34 points (with 18 contestants earning gold medals), the silver medal cutoff was 26 points (with 41 contestants earning silver medals), and the bronze medal cutoff was 17 points (with 48 contestants earning bronze medals).

In this IMO, only three contestants achieved a perfect score of 42 points, namely Vladimir Roganov and Stanislav Smirnov from the Soviet Union, and Géza Kós from Hungary.

Problem 5.24 (IMO 38-1, proposed by Belarus). In a plane points with integer coordinates are the vertices of unit squares. The squares are colored alternately black and white (as on a chessboard).

For any pair of positive integers m and n, consider a right-angled triangle whose vertices have integer coordinates and whose legs, of lengths m and n, lie along edges of the squares.

Let S_1 be the total area of the black part of the triangle and S_2 be the total area of the white part. Let $f(m, n) = |S_1 - S_2|$.

(a) Calculate $f(m, n)$ for all positive integers m and n which are either both even or both odd.
(b) Prove that $f(m, n) \le \frac{1}{2} \max\{m, n\}$ for all m and n.
(c) Show that there is no constant C such that $f(m, n) < C$ for all m and n.

Solution. (a) Let $\triangle ABC$ be one of the right triangles in question, with legs of length $AB = m$ and $AC = n$. Complete the rectangle $ABCD$ as shown in the Figure 5.14.

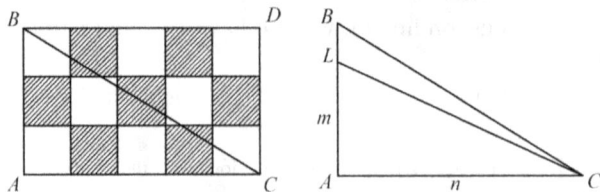

Figure 5.14

For any polygon P, denote by $S_1(P)$ the area of the black part of P and by $S_2(P)$ the area of the white part.

When m and n are both even or both odd, the coloring of the rectangle is symmetric with respect to the center of the rectangle. Thus

$$f(m,n) = |S_1(\triangle ABC) - S_2(\triangle ABC)|$$

$$= \frac{1}{2}|S_1(ABCD) - S_2(ABCD)|.$$

It follows that $f(m,n) = 0$ if m and n are both even; $f(m,n) = \frac{1}{2}$ if m and n are both odd.

(b) For m and n of the same parity, the value of $f(m,n)$ is 0 or $\frac{1}{2}$, and there is nothing to prove. We now inspect the remaining case. Assume that, e.g., m is odd and n is even. Take point L on side AB such that $AL = m - 1$. Since $m - 1$ is even, we have $f(m-1,n) = 0$, i.e., $S_1(\triangle ALC) = S_2(\triangle ALC)$. Therefore,

$$f(m,n) = |S_1(\triangle ABC) - S_2(\triangle ABC)|$$

$$= |S_1(\triangle LBC) - S_2(\triangle LBC)|$$

$$\leq S_1(\triangle LBC) + S_2(\triangle LBC) = \frac{n}{2} \leq \frac{1}{2}\max\{m,n\}.$$

(c) We now calculate the value of $f(2k+1, 2k)$. As shown in the Figure 5.15, $AB = 2k+1$ and $AC = 2k$. Let L be a point on AB such that $AL = 2k$. Since $f(2k, 2k) = 0$, we have $S_1(\triangle ALC) = S_2(\triangle ALC)$. Hence

$$f(2k+1, 2k) = |S_1(\triangle LBC) - S_2(\triangle LBC)|.$$

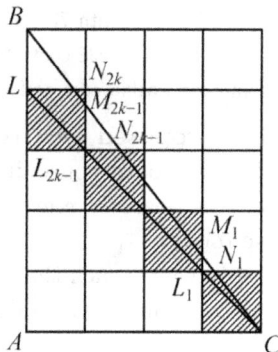

Figure 5.15

The area of $\triangle LBC$ is k. We may assume without loss of generality that the segment LC lies in the black part. Thus the white part of $\triangle LBC$ consists of $2k$ small triangles: $\triangle BLN_{2k}$, $\triangle M_{2k-1}L_{2k-1}N_{2k-1}, \ldots, \triangle M_1L_1N_1$, each of which is similar to $\triangle BAC$. Therefore, the total white area is

$$S_2(\triangle LBC) = \frac{1}{2} \cdot \frac{2k}{2k+1} \left(\left(\frac{2k}{2k}\right)^2 + \left(\frac{2k-1}{2k}\right)^2 + \cdots + \left(\frac{1}{2k}\right)^2 \right)$$

$$= \frac{1}{4k(2k+1)}(1^2 + 2^2 + \cdots + (2k)^2)$$

$$= \frac{4k+1}{12}.$$

And the black area is

$$S_1(\triangle LBC) = k - \frac{1}{12}(4k+1) = \frac{1}{12}(8k-1).$$

Thus, $f(2k+1, 2k) = \frac{2k-1}{6}$, which can be arbitrarily large.

【Score Situation】This particular problem saw the following distribution of scores among contestants: 48 contestants scored 7 points, 8 contestants scored 6 points, 15 contestants scored 5 points, 102 contestants scored 4 points, 23 contestants scored 3 points, 30 contestants scored 2 points, 143 contestants scored 1 point, and 91 contestants scored 0 point. The average score of this problem is 2.476, indicating that it had a certain level of difficulty.

Among the top five teams in the team scores, the scores of this problem are as follows: the China team scored 39 points (with a total team score of 223 points), the Hungary team scored 33 points (with a total team score of 219 points), the Iran team scored 33 points (with a total team score of 217 points), the Russia team scored 38 points (with a total team score of 202 points), and the United States team scored 37 points (with a total team score of 202 points).

The gold medal cutoff for this IMO was set at 35 points (with 39 contestants earning gold medals), the silver medal cutoff was 25 points (with 70 contestants earning silver medals), and the bronze medal cutoff was 15 points (with 122 contestants earning bronze medals).

In this IMO, a total of four contestants achieved a perfect score of 42 points.

Problem 5.25 (IMO 55-2, proposed by Croatia). Let $n \geq 2$ be an integer. Consider an $n \times n$ chessboard consisting of n^2 unit squares. A configuration of n rooks on this board is *peaceful* if every row and every column contains exactly one rook.

Find the greatest positive integer k such that for each peaceful configuration of n rooks, there is a $k \times k$ square which does not contain a rook on any of its k^2 unit squares.

Solution. We shall show that the answer is $k_{\max} = \lceil \sqrt{n-1} \rceil$ by two steps. Let l be a positive integer.

(1) If $n > l^2$, then there exists an empty $l \times l$ square for any peaceful configuration.

(2) If $n \leq l^2$, then there exists a peaceful configuration, such that each $l \times l$ square is not empty.

Proof of (1). There is a row R with the first column having a rook. Take successively l rows containing row R, which are denoted by U. If $n > l^2$, then $l^2 + 1 \leq n$, and from column 2 to column $l^2 + 1$ in U, containing the l number of $l \times l$ squares which have at most $(l - 1)$ rooks. So, at least one $l \times l$ square is empty.

Proof of (2). For $n = l^2$, we shall find a peaceful configuration which has no empty $l \times l$ square. We label the rows from bottom to top and the columns from left to right both by $0, 1, 2, \ldots, l^2 - 1$. So, denote by (r, c) the unit square at row r and column c.

We put a rook at $(il + j, jl + i)$ for $i, j = 0, 1, 2, \ldots, l - 1$. Figure 5.16 shows the case for $l = 3$. Since each number between 0 to $l^2 - 1$ can be written uniquely in the form of $il + j(0 \leq i, j \leq l - 1)$, such a configuration is peaceful.

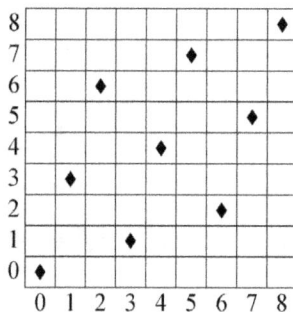

Figure 5.16

For any $l \times l$ square A, suppose that the lowest row of A is row $pl + q$ with $0 \leq p, q \leq l - 1$ (since $pl + q \leq l^2 - l$). There is one rook in A by the configuration, and the column labels of the rook may be $ql + p, (q + 1)l + p, \ldots, (l - 1)l + p, p + 1, l + (p + 1), \ldots, (q - 1)l + p + 1$. Rearrange these numbers in increasing order to

$$p + 1, l + (p + 1), \ldots, (q - 1)l + p + 1, \quad ql + p, (q + 1)l + p, \ldots, (l - 1)l + p.$$

Then the first number is less than or equal to l, the last is greater than or equal to $(l - 1)l$, and the difference between two adjacent numbers is l. Therefore, there exists one rook in A.

For the case of $n < l^2$, consider the configuration above, but delete $l^2 - n$ columns and rows from the right and from the bottom, respectively. We get an $l \times l$ square. Thus, we obtain an $n \times n$ square, where there is no empty $l \times l$ square. But some rows and columns may be empty. We can put rooks at the cross squares of empty rows and columns to obtain the desired peaceful configuration.

Remark. The answer could also be in the form of $[\sqrt{n}] - 1$.

【Score Situation】This particular problem saw the following distribution of scores among contestants: 122 contestants scored 7 points, 71 contestants scored 6 points, 39 contestants scored 5 points, 14 contestants scored 4 points, 17 contestants scored 3 points, 25 contestants scored 2 points, 32 contestants scored 1 point, and 240 contestants scored 0 point. The average score of this problem is 2.971, indicating that it had a certain level of difficulty.

Among the top five teams in the team scores, the scores of this problem are as follows: the China team scored 42 points (with a total team score of 201 points), the United States team scored 40 points (with a total team score of 193 points), the Chinese Taiwan team scored 38 points (with a total team score of 192 points), the Russia team scored 40 points (with a total team score of 191 points), and the Japan team scored 36 points (with a total team score of 177 points).

The gold medal cutoff for this IMO was set at 29 points (with 49 contestants earning gold medals), the silver medal cutoff was 22 points (with 113 contestants earning silver medals), and the bronze medal cutoff was 16 points (with 133 contestants earning bronze medals).

In this IMO, only three contestants achieved a perfect score of 42 points, namely Jiyang Gao from China, Po-Sheng Wu from Chinese Taiwan, and Alexander Gunning from Australia.

Problem 5.26 (IMO 57-2, proposed by Australia). Find all positive integers n for which each cell of an $n \times n$ table can be filled with one of the letters I, M, and O in such a way that:

(1) in each row and each column, one third of the entries are I, one third are M, and one third are O; and

(2) in any diagonal, if the number of entries on the diagonal is a multiple of three, then one third of the entries are I, one third are M, and one third are O.

Note. The rows and columns of an $n \times n$ table are each labeled 1 to n in a natural order. Thus, each cell corresponds to a pair of positive integers (i, j) with $1 \le i, j \le n$. For $n > 1$, the table has $4n - 2$ *diagonals* of two types. A diagonal of the first type consists of all cells (i, j) for which $i + j$

is a constant, and a diagonal of the second type consists of all cells (i, j) for which $i - j$ is a constant.

Solution. The answer is all the multiples of 9.

We first give a 9×9 table as follows:

$$
\begin{pmatrix}
I & I & I & M & M & M & O & O & O \\
M & M & M & O & O & O & I & I & I \\
O & O & O & I & I & I & M & M & M \\
I & I & I & M & M & M & O & O & O \\
M & M & M & O & O & O & I & I & I \\
O & O & O & I & I & I & M & M & M \\
I & I & I & M & M & M & O & O & O \\
M & M & M & O & O & O & I & I & I \\
O & O & O & I & I & I & M & M & M
\end{pmatrix}.
$$

It is easy to verify that this table satisfies the requirement.

For $n = 9k$, where k is a positive integer, we first divide an $n \times n$ table into k^2 subtables of size 9×9 and fill each subtable with the above pattern. For the $n \times n$ table, each of its row, or column, or diagonal where the number of entries is a multiple of 3 intersects every subtable at a row, a column, or a diagonal with the number of entries a multiple of 3 or 0 cells in the subtable; so it has the same number of I, M, and O's.

Next, we assume that an $n \times n$ table can be filled as required, and prove that $9 \mid n$. Since each row has the same number of I, M, and O's, so $3 \mid n$, and we denote $n = 3k$, where k is a positive integer. Divide the table into the k^2 number of 3×3 subtables; call the cell in the center of each subtable its *key cell*, and call a line — a row, a column, or a diagonal — passing through a key cell a *key line*. Let S be the set of all pairs (l, c), where l is a key line and c is a cell on l that is filled with an M. We count the size of S in two ways.

On the one hand, for each key line, exactly one third of its cells are filled with M. If l is one of the k key rows or k key columns, there are exactly k cells on l filled with M. The key diagonals of each type has $3, 6, 9, \ldots, 3k, 3k - 3, \ldots, 3$ cells on them, so

$$|S| = 2k \cdot k + 2 \times (1 + 2 + \cdots + k + (k - 1) + \cdots + 1) = 2k^2 + 2k^2 = 4k^2.$$

On the other hand, for each cell c, it is incident to exactly 1 key line, if it is not a key cell; or 4 key lines, if it is a key cell. There are $3k^2$ cells filled with M, and since $1 \equiv 4(\mathrm{mod}\,3)$, so

$$|S| \equiv 3k^2 \ (\mathrm{mod}\,3).$$

Therefore, $4k^2 = |S| \equiv 3k^2 \ (\mathrm{mod}\,3)$, so $3\,|\,k$ and $9\,|\,n$.

【Score Situation】 This particular problem saw the following distribution of scores among contestants: 107 contestants scored 7 points, 9 contestants scored 6 points, 8 contestants scored 5 points, 7 contestants scored 4 points, 30 contestants scored 3 points, 99 contestants scored 2 points, 65 contestants scored 1 point, and 277 contestants scored 0 point. The average score of this problem is 2.033, indicating that it had a certain level of difficulty.

Among the top five teams in the team scores, the scores of this problem are as follows: the United States team scored 42 points (with a total team score of 214 points), the South Korea team scored 41 points (with a total team score of 207 points), the China team scored 30 points (with a total team score of 204 points), the Singapore team scored 42 points (with a total team score of 196 points), and the Chinese Taiwan team scored 37 points (with a total team score of 175 points).

The gold medal cutoff for this IMO was set at 29 points (with 44 contestants earning gold medals), the silver medal cutoff was 22 points (with 101 contestants earning silver medals), and the bronze medal cutoff was 16 points (with 135 contestants earning bronze medals).

In this IMO, a total of six contestants achieved a perfect score of 42 points.

Problem 5.27 (IMO 63-6, proposed by Austria). Let n be a positive integer. A *Nordic square* is an $n \times n$ board containing all the integers from 1 to n^2 so that each cell contains exactly one number. Two different cells are considered adjacent if they share a common side. Every cell that is adjacent only to cells containing larger numbers is called a *valley*. An *uphill path* is a sequence of one or more cells such that:

(i) the first cell in the sequence is a valley;
(ii) each subsequent cell in the sequence is adjacent to the previous cell;
(iii) the numbers written in the cells in the sequence are in increasing order.

Find, as a function of n, the smallest possible total number of uphill paths in a Nordic square.

Solution. The smallest possible number of uphill paths is $2n^2 - 2n + 1$.

Let table A be the original $n \times n$ board. Define table B as another $n \times n$ board, and in each cell of B we write the number of uphill paths ending at that cell. The total number of uphill paths in A is equal to the sum of all entries of B: call this number S.

There is at least one valley in A, namely the cell with number 1. Each valley in A corresponds to 1 in B, and vice versa. Let T be the sum of all non-valley entries of B. Consider a pair of neighboring cells whose numbers are x and y with $x < y$: they contribute at least 1 to T, as there is always an uphill path through x and ending at y. Therefore, $T \geq 2n^2 - 2n$, the number of pairs of neighboring cells, and

$$S = T + \text{the number of valleys} \geq 2n^2 - 2n + 1.$$

It remains to prove that the lower bound $2n^2 - 2n + 1$ is attainable. To this end, we label some cells of B with dots (they will correspond to valleys in A): the dotted cells form a tree in B while the unlabeled cells are all isolated. Then, construct the table A as follows:

First, choose any dotted cell and write 1; then, consecutively write $2, 3, \ldots$ in the dotted cells such that a new number is always written next to an already written number; finally, write the remaining large numbers arbitrarily in the unlabeled cells.

Clearly, there is only one cell with number 1, and for each pair of neighboring cells, the one with the larger number is the end of exactly one uphill path. So, the number of uphill paths is indeed $2n^2 - 2n + 1$.

We label the cells in detail as follows: for the 1×1 table, label the unique cell; for the 2×2 table, label any three cells; for $n \times n (n \geq 3)$ tables, define

$$s = \begin{cases} 2, & \text{if } n \equiv 0, 2 \pmod 3, \\ 1, & \text{if } n \equiv 1 \pmod 3. \end{cases}$$

In the first column, label $(1, i)$ as long as $i \neq 6k + s$ (here, $(1, 1)$ is the lower left cell and $(1, n)$ is the upper left cell). In the second column, label $(2, j)$ if $j = 6k + s - 1$, $6k + s$, or $6k + s + 1$. Evidently, the dotted cells in the first two columns are connected.

Now we expand the dotted cells $(2, 6k + s)$ and $(1, 6k + s + 3)$ to the right: for each dotted (i, j), label all $(i + l, j)$ and $(i + 2l, j \pm 1)$ where l is any positive integer (we only label cells within the table). It is easy to see that the dotted cells form a tree and the unlabeled cells $(1, 6k + s)$, $(2 + 2l + 1, 6k + s \pm 1)$, and $(2 + 2l, 6k + s + 3 \pm 1)$ are all isolated (since, if they are in the same column, then the distance is at least 2; if they are in adjacent columns, then they are in different rows).

The following Figure 5.17 illustrates the labeling for $n = 3, 4, 5, 6, 7$ and Figure 5.18 gives tables A and B for $n = 5$ (there are other tables that take the same minimum value).

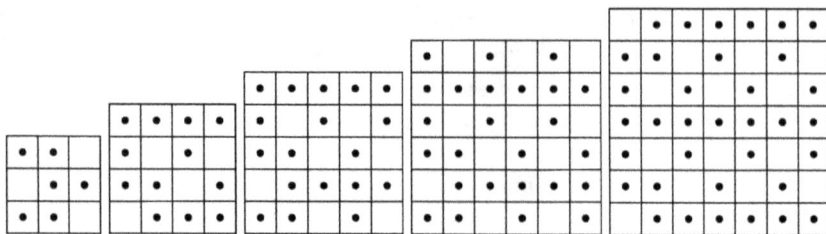

Figure 5.17 Labeling for $n = 3, 4, 5, 6, 7$

12	13	14	15	16
11	24	17	25	18
10	9	22	8	23
21	3	4	5	6
1	2	19	7	20

1	1	1	1	1
1	4	1	4	1
1	1	4	1	3
3	1	1	1	1
1	1	3	1	2

Figure 5.18 Tables A and B for $n = 5$

【Score Situation】 This particular problem saw the following distribution of scores among contestants: 22 contestants scored 7 points, 2 contestants scored 6 points, 3 contestants scored 5 points, 19 contestants scored 4 points, 7 contestants scored 3 points, 22 contestants scored 2 points, 80 contestants scored 1 point, and 434 contestants scored 0 point. The average score of this problem is 0.683, indicating that it was extremely difficult.

Among the top five teams in the team scores, the scores of this problem are as follows: the China team scored 42 points (with a total team score of 252 points), the South Korea team scored 15 points (with a total team score of 208 points), the United States team scored 26 points (with a total team score of 207 points), the Vietnam team scored 15 points (with a total team score of 196 points), and the Romania team scored 10 points (with a total team score of 194 points).

The gold medal cutoff for this IMO was set at 34 points (with 44 contestants earning gold medals), the silver medal cutoff was 29 points (with 101 contestants earning silver medals), and the bronze medal cutoff was 23 points (with 140 contestants earning bronze medals).

In this IMO, a total of 10 contestants achieved a perfect score of 42 points.

5.3 Summary

In the first 64 IMOs, there were a total of 27 combinatorial geometry problems. These problems can be broadly categorized into four types, as depicted in Figure 5.19. The score details for these problems are presented in Table 5.2. Due to the smaller number of participating teams and missing contestant score information in early IMOs, there are several blanks in Table 5.2.

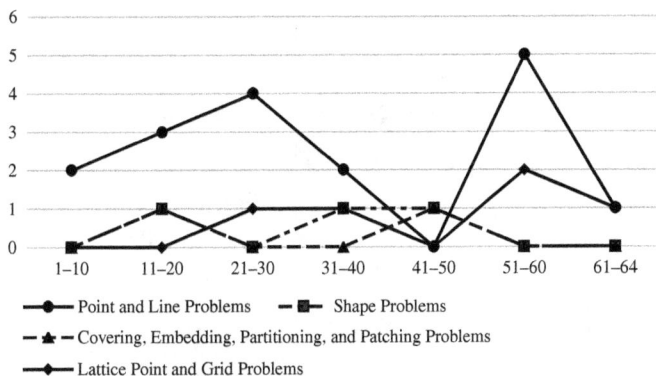

Figure 5.19 Numbers of Combinatorial Geometry Problems in the First 64 IMOs

Problems 5.1–5.17 focus on "point and line problems;" among these 17 problems, the one with the lowest average score is Problem 5.17 (IMO 61-6), proposed by Chinese Taiwan. Problems 5.18–5.20 deal with "shape problems;" among these 3 problems, the one with the lowest average score is Problem 5.19 (IMO 31-6), proposed by the Netherlands. Problems 5.21 to 5.22 are about "covering, embedding, partitioning, and patching problems;" among these two problems, the one with the lowest average score is Problem 5.22 (IMO 45-3), proposed by Estonia. Problems 5.23–5.27 are about "lattice point and grid problems;" among these five problems, the one with the lowest average score is Problem 5.27 (IMO 63-6), proposed by Austria.

These 27 problems were proposed by 19 countries and regions. The Soviet Union, Romania, the Netherlands, Bulgaria, Estonia, the German Democratic Republic, Australia, and Austria each contributed two problems.

From Table 5.2, it can be observed that in the first 64 IMOs, there were five combinatorial geometry problems with an average score of 0–1 point; six problems with an average score of 1–2 points; eight problems with an average score of 2–3 points; two problems with an average score of 3–4 points; six problems with an average score above 4 points.

Table 5.2 Score Details of Combinatorial Geometry Problems in the First 64 IMOs

Problem	5.1	5.2	5.3	5.4	5.5	5.6	5.7	5.8
Full points	7	9	7	7	6	7	7	7
Average score	3.647	2.750	4.179	1.393	2.813	2.555	4.050	1.583
Top five mean	4.300	3.950	5.825	4.225	4.350	2.425	6.033	3.300
6th–15th mean				0.657	2.456	3.425		2.117
16th–25th mean						1.500		0.483
Problem number in the IMO	6-5	7-6	11-5	13-5	17-5	23-6	24-4	25-3
Proposing country	Romania	Poland	Mongolia	Bulgaria	The Soviet Union	Vietnam	Belgium	Romania

Problem	5.9	5.10	5.11	5.12	5.13	5.14	5.15	5.16
Full points	7	7	7	7	7	7	7	7
Average score	4.249	3.126	4.298	0.654	2.526	0.339	4.307	0.806
Top five mean	6.800	6.800	6.633	2.600	5.733	2.867	6.767	4.000
6th–15th mean	5.967	5.267	6.400	1.091	5.133	1.217	6.433	2.197
16th–25th mean	4.017	4.383	5.242	1.030	3.750	0.233	6.217	1.182
Problem number in the IMO	28-5	36-3	40-1	52-2	54-2	55-6	56-1	57-6
Proposing country	East Germany	Japan	Estonia	The United Kingdom	Australia	Austria	The Netherlands	Czech

(Continued)

Table 5.2 (*Continued*)

Problem	5.17	5.18	5.19	5.20	5.21	5.22	5.23	5.24
Full points	7	8	7	7	6	7	7	7
Average score	0.282	2.400	1.503	1.833	5.064	1.012	1.948	2.476
Top five mean	3.700		3.500	5.433	5.625	3.367	4.933	6.000
6th–15th mean	0.867		2.409	4.015	5.260	2.067	2.900	4.250
16th–25th mean	0.617		1.463	2.759		1.583	1.086	3.150
Problem number in the IMO	61-6	12-6	31-6	47-2	16-4	45-3	27-6	38-1
Proposing country	Chinese Taiwan	The Soviet Union	The Netherlands	Serbia	Bulgaria	Estonia	East Germany	Belarus

Problem	5.25	5.26	5.27
Full points	7	7	7
Average score	2.971	2.033	0.683
Top five mean	6.533	6.400	3.600
6th–15th mean	5.867	4.833	1.867
16th–25th mean	5.250	3.424	1.067
Problem number in the IMO	55-2	57-2	63-6
Proposing country	Croatia	Australia	Austria

Note. Top five mean = Total score of the top five teams ÷ Total number of contestants from the top five teams,
6th–15th mean = Total score of the 6th–15th teams ÷ Total number of contestants from the 6th–15th teams,
16th–25th mean = Total score of the 16th–25th teams ÷ Total number of contestants from the 16th–25th teams.

Table 5.3 Numbers of Combinatorial Geometry Problems in the 24th–64th IMOs

Combinatorial Geometry Problem	Problem Number			Number of Problems in the First 64 IMOs
	1, 4	2, 5	3, 6	
Point and line problems	3	3	5	17
Shape problems	0	1	1	3
Covering, embedding, partitioning, and patching problems	0	0	1	2
Lattice point and grid problems	1	2	2	5
Total	4	6	9	27

In the 24th–64th IMOs, there were a total of 19 combinatorial geometry problems. Among these, five had an average score of 0–1 point; five had an average score of 1–2 points; four had an average score of 2–3 points; one had an average score of 3–4 points; four had an average score above 4 points.

Further analysis of the problem numbers of these 19 combinatorial geometry problems, as shown in Table 5.3, reveals that these problems frequently appeared as the 3rd/6th problem. The majority of these problems, totaling 11, were of the type of point and line problems. The other three types of combinatorial geometry problems were less frequent.

From Table 5.2, it is observable that in the combinatorial geometry problems, the average score of the top five teams generally exceeds the average score of the problem by 2 points, the average score of the 6th–15th teams typically surpasses the average score by 1.5 points, and the average score of the 16th–25th teams usually exceeds the average score by 1 point.

It can also be observed that there are several combinatorial geometry problems where the average score is higher than the average score of the 16th–25th teams, such as Problem 5.6 (IMO 23-6), Problem 5.8 (IMO 25-3), Problem 5.9 (IMO 28-5), Problem 5.19 (IMO 31-6), and Problem 5.23 (IMO 27-6).

This phenomenon is due to the smaller number of participating teams in early IMOs. It was not until the 30th IMO in 1989 that the number of participating teams exceeded 50. Therefore, it is common to see situations where the average score is close to or even higher than the average score of the 16th–25th teams during this period.

Chapter 6

Graph Theory Problems

Graph theory is a significant branch of modern mathematics that represents specific relationships in real-world entities through connections between points and lines. It plays an important role in fields such as operations research and computer science. The famous 18th-century problem of the Seven Bridges of Königsberg was solved by the mathematician Euler, which laid the foundations of graph theory and prefigured the idea of topology.

In mathematics competition problems with real-world contexts, such as outcomes of sports matches, friendship networks, or electrical circuits, graph theory often provides effective solutions.

In the first 64 IMOs, there had been a total of eight graph theory problems, accounting for approximately 9.2% of all combinatorics problems. The statistical distribution of these problems in the previous IMOs is presented in Table 6.1.

Table 6.1 Numbers of Graph Theory Problems in the First 64 IMOs

	Session							
Content	1–10	11–20	21–30	31–40	41–50	51–60	61–64	Total
Graph theory problems	1	0	1	3	0	1	2	8
Combinatorics problems	7	11	16	16	11	19	7	87
Percentage of graph theory problems among combinatorics problems	14.3%	0.0%	6.3%	18.8%	0.0%	5.3%	28.6%	9.2%

It is important to note that for each problem, the solutions are followed by information on the scores, including the number of contestants in each score range, the average score, and the scores of the top five teams. However, early IMOs often lacked information on contestant scores, so the number of contestants in each score range only represents the counted number of contestants, and some problems lack scores of the top five teams.

6.1   Common Theorems, Formulas, and Methods

6.1.1   *Preliminaries*

A graph is composed of several different vertices and edges connecting some of them. We usually denote a graph by G, the vertex set by V, and the edge set by E, and use the notation $G(V, E)$. For a graph G, the number of vertices $|V|$ is called the order of G. When $|V|$ and $|E|$ are both finite, we say that the graph is finite.

Unless otherwise specified, all graphs considered here are undirected simple graphs (that is the graph contains no loops nor parallel edges). If two vertices are connected by an edge, they are called adjacent; otherwise non-adjacent.

In a simple graph, the number of edges emanating from a vertex v is called the degree of this vertex, and denoted by $\deg v$. If $\deg v = 0$, then the vertex v is called isolated. If $\deg v = 1$, then the vertex v is called a leaf.

Vertices with odd degrees are called odd vertices; otherwise even vertices. If in a graph any two points are connected by an edge, it is called a complete graph. A complete graph with n vertices is denoted by K_n, and in this case, the number of edges is $|E| = \frac{1}{2}n(n-1)$.

If every vertex in a graph has degree r, then G is called an r-regular graph and denoted by $\deg G = r$ in this situation.

Clearly, if G is the complete graph K_n, then $\deg G = n - 1$.

Theorem 6.1 (Euler's Theorem). *In any graph $G = (V, E)$, the sum of all degrees is equal to twice the number of edges.*

In view of this theorem, we have several easy corollaries.

(1) The number of odd vertices is even.
(2) There exists a vertex whose degree is $\geq$ (or $\leq$) $\frac{2|E|}{|V|}$.

In a lot of cases, it is useful to consider a vertex with a maximum degree.

In a graph G, a sequence of distinct edges $e_1, e_2, \ldots, e_m$, satisfying $e_i = (v_{i-1}, v_i)$ for $i = 1, 2, \ldots, m$ with $v_0, v_1, \ldots, v_n$ distinct, is called a *path* from v_0 to v_m; m is the length of this path, and v_0 and v_m are called the endpoints of the path. This path may also be denoted as $v_0 v_1 \ldots v_m$.

The shortest length of paths between two vertices u and v is called the *distance* between them, denoted by $d(u, v)$. Clearly, the distance satisfies nonnegativity, symmetry, and the triangle inequality.

The maximum distance between the vertices of a graph is called the *diameter* of this graph. A complete graph has diameter 1.

If there exists a path between any two vertices in a graph, this graph is called connected. A maximal connected subgraph is called a connected component.

A walk in a graph is a sequence of edges which joins a sequence of vertices. A trail is a walk in which all edges are distinct. If $v_0 v_1 \cdots v_m$ is a trail such that v_0 and v_m coincide and all other vertices are distinct, then it is called a cycle; m is called the length of this cycle.

A path (resp., cycle) which passes all vertices is called a Hamiltonian path (resp., cycle).

Theorem 6.2. *A graph with the minimum degree ≥ 2 must contain a cycle.*

For the existence of a Hamiltonian path (or cycle), so far, we don't know a simple necessary and sufficient condition.

Theorem 6.3. *Let G be a simple graph of order $n(n \geq 3)$. If $\deg(u) + \deg(v) \geq n - 1$ for any two nonadjacent vertices u and v, then G contains a Hamiltonian path.*

Theorem 6.4 (Ore's Theorem). *Let G be a simple graph of order $n(n \geq 3)$. If $\deg(u) + \deg(v) \geq n$ for any two nonadjacent vertices u and v, then G contains a Hamiltonian cycle.*

Theorem 6.5 (Posa's Theorem). *Let G be a simple graph of order $n(n \geq 3)$. Assume that for every integer m with $1 \leq m < \frac{1}{2}(n-1)$, the number of vertices with degree at most m is less than m, and if n is odd and also the number of vertices with degree at most $\frac{1}{2}(n-1)$ is less than or equal to $\frac{1}{2}(n-1)$, then G contains a Hamiltonian cycle.*

A connected graph without cycles is called a *tree*. A graph whose connected components are trees is called a *forest*.

Theorem 6.6. *The number of vertices in a tree is equal to the number of edges plus 1. If there are at least two vertices, then there are at least two leaves.*

Trees can be considered as the simplest connected graphs. Removing one edge from a tree will make it disconnected and adding one edge to a tree will cause a cycle. Thus, trees are a good starting point for studying connected graphs.

Theorem 6.7. *A graph is a tree if and only if there exists a unique path between any two vertices.*

If there exists a trail in a graph which passes each edge exactly once, then G is called an *Eulerian graph.*

Theorem 6.8 (Euler's Theorem). *A finite graph is Eulerian if and only if it is connected and the number of odd vertices is either 0 or 2. If the number of odd vertices is 0, then the Eulerian path is closed.*

If the number of odd vertices is $2k$, then the edges can be covered by k paths.

Planar graph is also an important topic.

If a graph can be drawn in a plane such that the interior of the edges are disjoint, then this figure is called a planar graph. A planar graph divides the plane into disjoint regions.

Theorem 6.9 (Euler's Theorem for Planar Graphs). *If a connected planar graph G has v vertices, e edges, and f faces, then $v - e + f = 1$ (if the unbounded face is counted, then $v - e + f = 2$).*

Theorem 6.10. *If a connected planar graph has $v(\geq 3)$ vertices and e edges, then $e \leq 3v - 6$.*

Two graphs G_1 and G_2 are called isomorphic, if they have the same number of vertices and there is a bijection $f : G_1 \to G_2$ between their sets of vertices, such that for any two vertices A and B of G_1, they are adjacent in G_1 if and only if $f(A)$ and $f(B)$ are adjacent in G_2.

If two simple graphs share the same vertex set and their edge sets are disjoint with the union the complete graph, then they are called complement to each other. If a graph is isomorphic to its complement, we call this graph a self-complement graph.

A well-known result about Ramsey's theorem is the following:

Theorem 6.11 (Ramsey's Theorem). *Among any six people in the world, there exist three of them, who either know each other or don't know each other.*

If each edge of K_n is colored by one of the k colors $c_1, c_2, \ldots, c_k$, it is called a k-colored K_n.

Theorem 6.12 (Goodman's Theorem). *A 2-colored K_6 contains at least two monochromatic triangles.*

Theorem 6.13. *A 3-colored K_{17} contains at least three monochromatic triangles.*

For sufficiently large n, a k-colored K_n contains a monochromatic triangle. The smallest such n is denoted by r_k, called the Ramsey numbers.

Theorem 6.14. (1) *For each integer $k \geq 2$, the Ramsey number r_k exists and satisfies the inequality $r_k \leq k(r_{k-1} - 1) + 2$;*
(2) *for any positive integer k,*

$$r_k \leq 1 + 1 + k + k(k-1) + \cdots + \frac{k!}{2!} + \frac{k!}{1!} + k!.$$

Assume that each edge of K_n is colored red or blue. For fixed positive integers p and q, when n is sufficiently large, there always exists either a red K_p subgraph, or a blue K_q subgraph. The smallest n with this property is denoted $r(p, q)$, which are also called the Ramsey numbers.

Theorem 6.15. $r(3,3) = 6$, $r(2,n) = n$, $r(3,4) = 9$, $r(3,5) = 14$, $r(3,6) = 18$, $r(3,7) = 23$, *and* $r(4,4) = 18$.

Theorem 6.16 (Erdös's Theorem). *The Ramsey numbers satisfy the following inequality:*

$$r(p,q) \leq r(p-1,q) + r(p,q-1).$$

Theorem 6.17 (Erdös-Szekeres, Greenwood-Gleason Theorem). *For $p \geq 2$ and $q \geq 2$,*

$$r(p,q) \leq C_{p+q-2}^{p-1}.$$

Theorem 6.18 (Schur's Theorem). *For any positive integer k, there exists n_0 such that for any $n \geq n_0$, any k-colored set $\{1, 2, \ldots, n\}$ contains monochromatic $x, y, z \in \{1, 2, \ldots, n\}$ (not necessarily distinct), satisfying $x + y = z$.*

6.1.2 *Common methods*

Many combinatorial problems may not seem like graph theory problems, but they can be transformed into graph theory problems, which may be more convenient. Then, we may apply the results in graphs theory, combining the methods of existence problems and combinatorial extremal problems.

Example 6.1. There are $n(\geq 4)$ contestants $A_1, A_2, \ldots, A_n$ participating in a math competition. Some students are friends, and any two students who are not friends have exactly two common friends. Assume that contestants

A_1 and A_2 are friends, and they don't have common friends. Prove that they have the same number of friends.

Proof. Construct a graph G with n vertices $v_1, v_2, \ldots, v_n$ representing the n contestants $A_1, A_2, \ldots, A_n$; draw an edge if these two contestants are friends. By assumption, the graph G satisfies the following property: any two non-adjacent vertices have exactly two common neighbors, and v_1 and v_2 are adjacent, but they have no common neighbors. We want to prove $d(v_1) = d(v_2)$. It suffices to prove $d(v_1) \leq d(v_2)$ and $d(v_2) \leq d(v_1)$.

Let M_A be the set of the vertices other than v_2 that are adjacent to v_1 and M_B be the set of the vertices other than v_1 that are adjacent to v_2. By assumption, M_A and M_B are disjoint. If $M_A = \varnothing$, then $d(v_1) = 1 \leq d(v_2)$; if $M_A \neq \varnothing$, then by choosing arbitrary $v_i \in M_A$, we have $v_i \notin M_B$ since M_A and M_B are disjoint. Therefore v_i and v_2 are not adjacent.

By assumption, there is a unique v_j other than v_1 which is a common neighbor of v_i and v_2. Hence $v_j \in M_B$ (As shown in Figure 6.1). The correspondence v_i to v_j is injective, for if there are two vertices v_i and v_k in M_A that correspond to the same v_j, then A_1 and A_j have 3 common friends, A_i, A_k, and A_2, a contradiction.

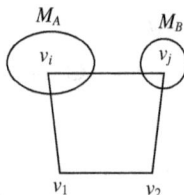

Figure 6.1

Thus $|M_A| \leq |M_B|$ and $d(v_1) = |M_A| + 1 \leq |M_B| + 1 = d(v_2)$. Similarly, we have $d(v_2) \leq d(v_1)$. Thus $d(v_1) = d(v_2)$, that is, A_1 and A_2 have the same number of friends.

6.2   Problems and Solutions

Problem 6.1 (IMO 6-4, proposed by Hungary). Seventeen people correspond by mail with one another — each one with all the rest. In their letters only three different topics are discussed. Each pair of correspondents deals with only one of these topics. Prove that there are at least three people who write to each other about the same topic.

Proof. Select any person, say A. He corresponds with 16 other people. Since there are only three topics, he must write to at least six of them on the same topic due to the pigeonhole principle. Assume for definiteness that A writes to six other people on topic I. If any one of these six writes to another of them on topic I, then there are 3 writers corresponding to topic I.

Assume that these six write to each other only on topics II and III. If B is one of the six, then by the pigeonhole principle, he must write to at least three of the other five on one of the two topics, say II. If two of these three correspond to each other on topic II, then we have found three people corresponding on topic II. If, on the other hand, all three write to each other on topic III, then the assertion also holds.

【Score Situation】 This particular problem saw the following distribution of scores among contestants: 7 contestants scored 6 points, no contestant scored 5 points, no contestant scored 4 points, no contestant scored 3 points, no contestant scored 2 points, no contestant scored 1 point, and 10 contestants scored 0 point. The average score of this problem is 2.471, indicating that it had a certain level of difficulty.

Among the top five teams in the team scores, the scores of this problem are as follows: the Soviet Union team scored 24 points (with a total team score of 269 points), the Hungary team scored 31 points (with a total team score of 253 points), the Romania team scored 2 points (with a total team score of 213 points), the Poland team scored 6 points (with a total team score of 209 points), and the Bulgaria team scored 18 points (with a total team score of 198 points).

The gold medal cutoff for this IMO was set at 38 points (with 7 contestants earning gold medals), the silver medal cutoff was 31 points (with 9 contestants earning silver medals), and the bronze medal cutoff was 27 points (with 19 contestants earning bronze medals).

In this IMO, only one contestant achieved a perfect score of 42 points.

Problem 6.2 (IMO 21-2, proposed by Bulgaria). A prism with pentagons $A_1A_2A_3A_4A_5$ and $B_1B_2B_3B_4B_5$ as top and bottom faces is given. Each side of the two pentagons and each of the line-segments A_iB_j for all $i, j = 1, 2, 3, 4, 5$ is colored either red or green. Every triangle whose vertices are vertices of the prism and whose sides have all been colored has two sides of a different color.

Show that all 10 sides of the top and bottom faces are of the same color.

Proof. We first show that the edges A_1A_2, A_2A_3, A_3A_4, A_4A_5, and A_5A_1 are all of the same color by an indirect proof. Suppose on the contrary that, say, edge A_1A_2 is red and A_2A_3 is green. At least three of the five segments A_2B_1, A_2B_2, A_2B_3, A_2B_4, and A_2B_5 have the same color.

Suppose without loss of generality that these are red, and label them A_2B_i, A_2B_j, and A_2B_k. Then at least one of the segments B_iB_j, B_jB_k, and B_kB_i is an edge of the base pentagon $B_1B_2B_3B_4B_5$; call it B_rB_s.

If B_rB_s is red, we would have a red triangle $A_2B_rB_s$.

Therefore, B_rB_s is green. Now segments A_1B_r and A_1B_s must also be green, for otherwise we would have $\triangle A_1A_2B_r$ or $\triangle A_1A_2B_s$ as red triangles. Therefore, $\triangle A_1B_rB_s$ is a green triangle. This contradiction implies that A_1A_2 and A_2A_3 have the same color and similarly that all the edges of each base have the same color.

Suppose on the contrary that the edges of $A_1A_2A_3A_4A_5$ are red and the edges of $B_1B_2B_3B_4B_5$ are green. If three of the 5 edges $A_1B_1, A_1B_2, \ldots, A_1B_5$ are green, say A_1B_i, A_1B_j, and A_1B_k, then at least one of B_iB_j, B_jB_k, and B_kB_i is an edge of the pentagon $B_1B_2B_3B_4B_5$; call it B_rB_s.

Consequently, then $A_1B_rB_s$ is a green triangle, a contradiction. Thus at least three of the five edges $A_1B_1, A_1B_2, \ldots, A_1B_5$ are red. Similarly, at least three of the five edges $A_2B_1, A_2B_2, \ldots, A_2B_5$ are red. Two of these six red edges must terminate on the same vertex B_i of the bottom. Then $\triangle A_1A_2B_i$ is a red triangle, a contradiction.

Thus, the edges of the two bases are all the same color.

【Score Situation】 This particular problem saw the following distribution of scores among contestants: 49 contestants scored 7 points, 3 contestants scored 6 points, 8 contestants scored 5 points, 11 contestants scored 4 points, 11 contestants scored 3 points, 10 contestants scored 2 points, 14 contestants scored 1 point, and 60 contestants scored 0 point. The average score of this problem is 3.084, indicating that it was relatively straightforward.

Among the top five teams in the team scores, the scores of this problem are as follows: the Soviet Union team scored 45 points (with a total team score of 267 points), the Romania team scored 34 points (with a total team score of 240 points), the Germany team scored 46 points (with a total team score of 235 points), the United Kingdom team scored 33 points (with a total team score of 218 points), and the United States team scored 40 points (with a total team score of 199 points).

The gold medal cutoff for this IMO was set at 37 points (with 8 contestants earning gold medals), the silver medal cutoff was 29 points (with 32 contestants earning silver medals), and the bronze medal cutoff was 20 points (with 42 contestants earning bronze medals).

In this IMO, a total of four contestants achieved a perfect score of 40 points.

Problem 6.3 (IMO 31-2, proposed by Czechoslovakia). Let $n \geq 3$ and consider a set E of $2n - 1$ distinct points on a circle. Suppose that exactly k of these points are to be colored black. Such a coloring is "good" if there is at least one pair of black points such that the interior of one of the arcs between them contains exactly n points from E.

Find the smallest value of k so that every such coloring of k points of E is good.

Solution. Label the $2n - 1$ points, in cyclic order, $v_1, v_2, \ldots, v_{2n-1}$. Construct a graph $G = (V, E)$, where $V = \{v_1, v_2, \ldots, v_{2n-1}\}$ and $E = \{v_i v_{i+n+1} \mid i = 1, 2, \ldots, 2n - 1\}$; here the indices are considered modulo $2n - 1$.

Clearly v_i is only adjacent to v_{i+n+1} and v_{i-n-1}. Since $v_{i+n+1} \neq v_{i-n-1}$, it follows that G is a 2-regular graph. Hence G is a union of disjoint cycles. Note that,

$$\gcd(n+1, 2n-1) = \begin{cases} 1, & n \equiv 0, 1 \pmod 3, \\ 3, & n \equiv 2 \pmod 3. \end{cases}$$

For $n \equiv 0, 1 \pmod 3$, the graph $G = C_{2n-1}$ is a cycle of length $2n - 1$. By assumption, there exist two adjacent vertices on the cycle among any k vertices. The minimum such k is $\left[\frac{2n-1}{2}\right] + 1 = n$. (On a cycle of length m, one can choose $\left[\frac{m}{2}\right]$ mutually non-adjacent vertices, and there exit two adjacent vertices among any $1 + \left[\frac{m}{2}\right]$ vertices.)

For $n \equiv 2 \pmod 3$, the graph G is a union of three cycles of length $\frac{2n-1}{3}$. The minimum value of k is $3\left[\frac{2n-1}{6}\right] + 1 = n - 1$.

In conclusion,

$$k_{\min} = \begin{cases} n, & 3 \nmid n - 2, \\ n - 1, & 3 \mid n - 2. \end{cases}$$

【Score Situation】 This particular problem saw the following distribution of scores among contestants: 96 contestants scored 7 points, 8 contestants scored 6 points, 9 contestants scored 5 points, 11 contestants scored 4 points, 54 contestants scored 3 points, 37 contestants scored 2 points, 46 contestants scored 1 point, and 47 contestants scored 0 point. The average score of this problem is 3.542, indicating that it was relatively straightforward.

Among the top five teams in the team scores, the scores of this problem are as follows: the China team scored 42 points (with a total team score of 230 points), the Soviet Union team scored 38 points (with a total team score of 193 points), the United States team scored

38 points (with a total team score of 174 points), the Romania team scored 32 points (with a total team score of 171 points), and the France team scored 34 points (with a total team score of 168 points).

The gold medal cutoff for this IMO was set at 34 points (with 23 contestants earning gold medals), the silver medal cutoff was 23 points (with 56 contestants earning silver medals), and the bronze medal cutoff was 16 points (with 76 contestants earning bronze medals).

In this IMO, a total of four contestants achieved a perfect score of 42 points.

Problem 6.4 (IMO 32-4, proposed by the United States). Suppose G is a connected graph with k edges. Prove that it is possible to label the edges $1, 2, \ldots, k$ in such a way that at each vertex which belongs to two or more edges, the greatest common divisor of the integers labeling those edges is equal to 1.

Proof. By the connectivity of G, each vertex of G has degree at least 1. Pick arbitrarily a vertex v_0 of G, and start a walk from v_0, moving along the edges of G. Each edge is allowed to travel only once; however each vertex can be visited multiple times. Assume that it cannot keep moving after travelling through l_1 edges, now label the vertices by $v_0, v_1, \ldots, v_{l_1}$ in order of their appearance on the route (v_i and v_j may be the same vertex for $0 \le i < j \le l_1$), and label the edges in order by $1, 2, \ldots, l_1$ with $1 \le l_1 \le k$.

Note that for any vertex v other than v_0, if two or more edges from v are labeled, then two of the edges are labeled consecutive integers. Thus the greatest common divisor of the integers labeling these edges is 1. If only one edge from v is labeled, then v is the terminal point of this route and it has degree 1, and thus it does not matter.

If $l_1 = k$, then we are done. Assume that $1 \le l_1 < k$. By the connectivity of G, there exists a vertex, either v_0 or the one that has at least two edges labeled, which has edges unlabeled.

We start from such a vertex and walk through the unlabeled edges until we cannot move anymore. Now label the edges we travelled by $l_1 + 1, \ldots, l_1 + l_2$ in the order they appear on the route, so $1 \le l_2 \le k - l_1$. We may continue this procedure until all edges are labeled $1, 2, \ldots, k$.

For any vertex v of G with degree ≥ 2, if $v = v_0$, then since one edge from v_0 is labeled 1, the greatest common divisor of all edges from v_0 is 1. If $v \ne v_0$, then we see from the construction that two of the edges from v are labeled consecutive integers; hence the greatest common divisor is also 1.

【Score Situation】 This particular problem saw the following distribution of scores among contestants: 95 contestants scored 7 points, 13 contestants scored 6 points, 15 contestants scored 5 points, 10 contestants scored 4 points, 18 contestants scored 3 points, 15 contestants scored 2 points, 34 contestants scored 1 point, and 112 contestants scored 0 point. The average score of this problem is 3.128, indicating that it was relatively straightforward.

Among the top five teams in the team scores, the scores of this problem are as follows: the Soviet Union team scored 34 points (with a total team score of 241 points), the China team scored 31 points (with a total team score of 231 points), the Romania team scored 27 points (with a total team score of 225 points), the Germany team scored 33 points (with a total team score of 222 points), and the United States team scored 20 points (with a total team score of 212 points).

The gold medal cutoff for this IMO was set at 39 points (with 20 contestants earning gold medals), the silver medal cutoff was 31 points (with 51 contestants earning silver medals), and the bronze medal cutoff was 19 points (with 84 contestants earning bronze medals).

In this IMO, a total of nine contestants achieved a perfect score of 42 points.

Problem 6.5 (IMO 33-3, proposed by China). Consider nine points in space, no four of which are coplanar. Each pair of points is joined by an edge (that is, a line segment) and each edge is either colored blue or red or left uncolored.

Find the smallest value of n such that whenever exactly n edges are colored, the set of colored edges necessarily contains a triangle whose all edges have the same color.

Solution. We first show that $n = 33$ has the prescribed property. Assume that there are 33 edges colored red or blue. Then there are $C_9^2 - 33 = 36 - 33 = 3$ left uncolored.

Let A_1 be a vertex such that it has an uncolored edge, we delete A_1 and all edges of A_1. If there is some vertex in the remaining graph, say A_2, which has an uncolored edge, delete A_2 and all edges of A_2, and keep doing this until the remaining graph does not have any uncolored edges.

Since there are three edges in the original graph, we have deleted at most three vertices, and there remains at least six vertices. It is a well-known Ramsey Theorem that a two-colored complete graph of order 6 has a monochromatic triangle.

The following example shows that there is not necessarily a monochromatic triangle if only 32 edges are colored. As shown in the Figure 6.2, the edges v_1v_9, v_2v_8, v_3v_7, and v_4v_6 are uncolored.

The solid lines and the dotted lines refer to the red edges and blue edges, respectively. For example, the dotted line between v_1, v_9 and v_2, v_8 refers

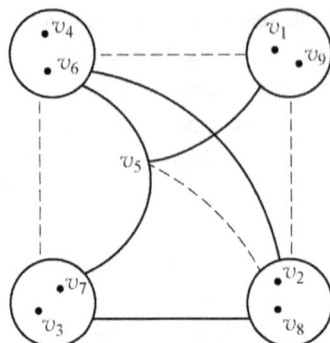

Figure 6.2

to four blue edges v_1v_2, v_1v_8, v_9v_2, and v_9v_8. It is now easy to verify that there is no triangle in this graph whose edges have the same color.

【Score Situation】 This particular problem saw the following distribution of scores among contestants: 59 contestants scored 7 points, 5 contestants scored 6 points, 7 contestants scored 5 points, 11 contestants scored 4 points, 17 contestants scored 3 points, 19 contestants scored 2 points, 55 contestants scored 1 point, and 177 contestants scored 0 point. The average score of this problem is 1.903, indicating that it was relatively challenging.

Among the top five teams in the team scores, the scores of this problem are as follows: the China team scored 42 points (with a total team score of 240 points), the United States team scored 32 points (with a total team score of 181 points), the Romania team scored 29 points (with a total team score of 177 points), the Commonwealth of Independent States team scored 35 points (with a total team score of 176 points), and the United Kingdom team scored 23 points (with a total team score of 168 points).

The gold medal cutoff for this IMO was set at 32 points (with 26 contestants earning gold medals), the silver medal cutoff was 24 points (with 55 contestants earning silver medals), and the bronze medal cutoff was 14 points (with 74 contestants earning bronze medals).

In this IMO, a total of four contestants achieved a perfect score of 42 points.

Problem 6.6 (IMO 60-3, proposed by Croatia). A social network has 2019 users, some pairs of whom are friends. Whenever user A is a friend toward user B, user B is also a friend toward user A. Events of the following kind may happen repeatedly, one at a time:

When three users A, B, and C such that A is a friend toward both B and C, but B and C are not friends, change their friendship statuses such that B and C are now friends, but A is no longer a friend toward B and C. All other friendship statuses are unchanged.

Initially, 1010 users have 1009 friends each, and 1009 users have 1010 friends each. Prove that there exists a sequence of such events after which each user is a friend of at most one other user.

Proof. We interpret this problem in the language of graph theory. First, there is a simple graph with 2019 vertices, in which 1010 vertices have degree 1009, and the other 1009 vertices have degree 1010.

The following operation is allowed: if A is adjacent to B and C, while B and C are not adjacent, then we delete edges AB and AC, and connect BC. We proceed to show that there is a sequence of operations such that after these operations, the graph has only isolated vertices and edges with no common vertices.

Note that for any two vertices u and v in G, the sum of their degrees is at least 2018, so it follows that either u and v are adjacent to each other or they have a common neighbor. Hence G is connected. Apparently, G is not a complete graph, and it contains vertices with odd degrees.

In fact, we shall show that the desired conclusion holds for every simple graph that is connected, non-complete, and contains a vertex with an odd degree. In the following argument we only assume that G has these properties.

Step 1: If G is not a tree, then we claim that there exists a sequence of operations, after which G becomes a tree. Note that an operation does not change the parity of the degree of any vertex (if a vertex has an odd degree, its degree remains odd after any sequence of operations, and vice versa).

Thus, after any sequence of operations G still has a vertex with an odd degree. It suffices to show that if G is not a tree, then there exists an operation that leaves G connected, so that since the number of edges decreases strictly, it will eventually become a tree after finitely many operations.

We choose the longest cycle in G, namely $x_1, x_2, \ldots, x_k, x_1$. If it is not a Hamiltonion cycle, then there is a vertex (which we assume to be x_1) that is adjacent to some vertex y outside the cycle. Then y is not adjacent to x_2, since otherwise $x_1, y, x_2, \ldots, x_k, x_1$ is a longer cycle. Then we delete edges $x_1 y$ and $x_1 x_2$, and then add $y x_2$. It is easy to see that the graph remains connected.

On the other hand, it is a Hamiltonion cycle, since there is a vertex with an odd degree and the graph is not complete, the graph must contain an edge, which we assume to be $x_1 x_i$, such that either x_1 is not adjacent to x_{i+1} or x_1 is not adjacent to x_{i-1}. Without loss of generality, we assume that x_1 is not adjacent to x_{i+1}. Then we may delete edges $x_1 x_i$ and $x_i x_{i+1}$ and add $x_1 x_{i+1}$, so the graph remains connected.

Step 2: If G is a tree, we claim that we may do a sequence of operations to change it into a graph with only isolated vertices and edges with no common vertices. If G has only 1 or 2 vertices, then the conclusion holds automatically.

Suppose G has at least three vertices. Then there exists a vertex u whose degree is at least 2. Let uv and uw be two edges incident to u. Then v is not adjacent to w.

Now we delete edges uv and uw, and then add vw. Then the graph is divided into two connected components, each of which is still a tree. For every component with at least three vertices, we continue with the above operation, until every component has at most two vertices. Finally, we obtain the desired configuration.

【Score Situation】 This particular problem saw the following distribution of scores among contestants: 28 contestants scored 7 points, 4 contestants scored 6 points, 9 contestants scored 5 points, 5 contestants scored 4 points, 6 contestants scored 3 points, 3 contestants scored 2 points, 46 contestants scored 1 point, and 520 contestants scored 0 point. The average score of this problem is 0.572, indicating that it was extremely difficult.

Among the top five teams in the team scores, the scores of this problem are as follows: the China team scored 27 points (with a total team score of 227 points), the United States team scored 26 points (with a total team score of 227 points), the South Korea team scored 26 points (with a total team score of 226 points), the North Korea team scored 17 points (with a total team score of 187 points), and the Thailand team scored 5 points (with a total team score of 185 points).

The gold medal cutoff for this IMO was set at 31 points (with 52 contestants earning gold medals), the silver medal cutoff was 24 points (with 94 contestants earning silver medals), and the bronze medal cutoff was 17 points (with 156 contestants earning bronze medals).

In this IMO, a total of six contestants achieved a perfect score of 42 points.

Problem 6.7 (IMO 61-3, proposed by Hungary). There are $4n$ pebbles of weights $1, 2, \ldots, 4n$. Each pebble is colored in one of n colors and there are four pebbles of each color. Show that we can arrange the pebbles into two piles so that the following two conditions are both satisfied:

- The total weights of both piles are the same.
- Each pile contains two pebbles of each color.

Proof. Let the n colors be $c_1, c_2, \ldots, c_n$, and let $\{1, 2, \ldots, 4n\}$ denote the set of pebbles, where i represents the pebble of weight i. We construct a graph $G = (V, E)$, where $V = \{c_1, c_2, \ldots, c_n\}$, and the set of edges is defined in the following way: if pebble i has color u, and pebble $4n + 1 - i$ has color v, then we connect u and v with an edge.

Thus, G has exactly $2n$ edges, which may contain loops (which occur when i and $4n + 1 - i$ have the same color) and multi-edges. Since every color has exactly 4 pebbles, the graph G is 4-regular (note that a loop is counted 2 degrees at the same vertex). We then prove the following lemma.

Lemma. *Every 4-regular graph (not necessarily simple) has a 2-regular spanning subgraph.*

Proof of Lemma. It suffices to prove the assertion for connected graphs, since in the general case we may take the union of the subgraphs in all the components. Let $G = (V, E)$ be 4-regular, with $|V| = n$ and $|E| = 2n$.

Then since every vertex has an even degree, this graph is Eulerian and there exists a Eulerian circuit L. Now we place the edges of L alternately into E_1 and E_2. Then $|E_1| = |E_2| = n$, and we claim that both graphs (V, E_1) and (V, E_2) are 2-regular.

In fact, for a vertex v, if there are no loops at v, then L passes through v twice, and every time the edge coming in and the edge going out belong to a different E_i, so exactly 2 of the edges incident to v belong to E_1 and the remaining 2 belong to E_2.

If there is one loop at v, then there are three consecutive edges e_1, e_2, and e_3, where e is a loop at v, and e_1, e_2 are two non-loop edges incident to v. By construction e_1, e_2 belong to the same E_i, and e belongs to the other part. Since e is counted 2 degrees, v still has degree 2 in (V, E_1) and (V, E_2).

If there are two loops at v, then since the graph is connected, the only possible case is $n = 1$. In this case, one of the two loops belongs to E_1 and the other belongs to E_2. Therefore, the same assertion still holds. Hence the lemma is proven.

Coming back to the original problem, since our graph $G = (V, E)$ is 4-regular, its edge set admits a partition E_1 and E_2 with $|E_1| = |E_2| = n$, and (V, E_1) and (V, E_2) are both 2-regular spanning subgraphs. By the definition of G, every edge corresponds to a pair $(i, 4n + 1 - i)$.

Now we put the n pairs of pebbles corresponding to the n edges in E_i into the i-th pile $(i = 1, 2)$. Since each pair has a total weight $4n + 1$, the total weight in each pile is exactly $n(4n + 1)$. Further, since (V, E_1) and (V, E_2) are 2-regular, each pile contains exactly 2 pebbles in each color.

Remark. The lemma we have proved is a special case of the theorem below. For a graph $G = (V, E)$, if every vertex has degree k, then we call G a k-regular graph. If a spanning subgraph H of G is r-regular, then we call H an r-factor of G. A 1-factor is also called a perfect match.

Theorem 6.19 (Petersen's 2-Factor Theorem). *If G is a $2k$-regular graph (that may contain loops and multi-edges), then G has a 2-factor.*

【Score Situation】This particular problem saw the following distribution of scores among contestants: 42 contestants scored 7 points, 5 contestants scored 6 points, no contestant scored 5 points, 40 contestants scored 4 points, 14 contestants scored 3 points, 3 contestants scored 2 points, 47 contestants scored 1 point, and 465 contestants scored 0 points. The average score of this problem is 0.940, indicating that it was extremely difficult.

Among the top five teams in the team scores, the scores of this problem are as follows: the China team scored 31 points (with a total team score of 215 points), the Russia team scored 28 points (with a total team score of 185 points), the United States team scored 15 points (with a total team score of 183 points), the South Korea team scored 22 points (with a total team score of 175 points), and the Thailand team scored 15 points (with a total team score of 174 points).

The gold medal cutoff for this IMO was set at 31 points (with 49 contestants earning gold medals), the silver medal cutoff was 24 points (with 112 contestants earning silver medals), and the bronze medal cutoff was 15 points (with 155 contestants earning bronze medals).

In this IMO, only one contestant achieved a perfect score of 42 points, namely Jinmin Li from China.

Problem 6.8 (IMO 61-4, proposed by India). There is an integer $n > 1$. There are n^2 stations on a slope of a mountain, all at different altitudes. Each of two cable car companies, A and B, operates k cable cars; each cable car provides a transfer from one of the stations to a higher one (with no intermediate stops).

The k cable cars of A have k different starting points and k different finishing points, and a cable car which starts higher also finishes higher. The same conditions hold for B. We say that two stations are *linked* by a company if one can start from the lower station and reach the higher one by using one or more cars of that company (no other movements between stations are allowed).

Determine the smallest positive integer k for which one can guarantee that there are two stations that are linked by both companies.

Solution. The answer is $n^2 - n + 1$.

We first show that if $k \leq n^2 - n$, then there exists a situation where no two stations are linked by both companies. Apparently, it suffices to give an example for $k = n^2 - n$, since for smaller k we simply delete some of the cable cars.

Let $S_1, S_2, \ldots, S_{n^2}$ be the stations with increasing altitudes. We set the $n^2 - n$ cable cars of A to be from S_i to S_{i+1}, where $1 \leq i \leq n^2 - 1$ and $n \nmid i$. Then set the $n^2 - n$ cable cars of B to be from S_i to S_{i+n}, where $1 \leq i \leq n^2 - n$.

Thus, two stations S_i and S_j are linked by A if and only if $\lceil \frac{i}{n} \rceil = \lceil \frac{j}{n} \rceil$, and they are linked by B if and only if $i \equiv j \pmod{n}$. These two conditions cannot be satisfied simultaneously, so no two stations are linked by both A and B.

Next, we prove that for $k = n^2 - n + 1$, such two stations always exist. Construct a directed graph G_A as follows: the vertex set consists of the n^2 stations $\{S_1, S_2, \ldots, S_{n^2}\}$, and for $1 \leq i < j \leq n^2$, if A runs a cable car from S_i to S_j, then there is a directed edge $S_i \rightarrow S_j$. It follows from the conditions that every vertex in G_A has at most one in-degree and at most one out-degree. Also, any edge in G_A points towards the higher station, so the graph does not contain any directed cycle.

It is easy to see that G_A consists of several directed chains (here a directed chain may consist of only one vertex), where two stations are connected by A if and only if they belong to the same directed chain. Since there are $n^2 - n + 1$ edges, it follows that exactly $n^2 - n + 1$ vertices have in-degree, which means exactly $n - 1$ vertices have no in-degree, so there are exactly $n - 1$ directed chains in G_A.

By the pigeonhole principle, some chain contains at least $\lceil \frac{n^2}{n-1} \rceil =$ $n + 2$ vertices and let X be the set of vertices in this chain. Then $|X| \geq n + 2$.

Construct G_B similarly, and by the same argument G_B also has $n - 1$ directed chains. Hence, there are two vertices S_i and S_j in X that belong to the same chain in G_B, so they are linked by both companies.

【Score Situation】 This particular problem saw the following distribution of scores among contestants: 285 contestants scored 7 points, 13 contestants scored 6 points, 14 contestants scored 5 points, 35 contestants scored 4 points, 42 contestants scored 3 points, 3 contestants scored 2 points, 11 contestants scored 1 point, and 213 contestants scored 0 point. The average score of this problem is 3.938, indicating that it was relatively straightforward.

Among the top five teams in the team scores, the scores of this problem are as follows: the China team scored 42 points (with a total team score of 215 points), the Russia team scored 42 points (with a total team score of 185 points), the United States team scored 42 points (with a total team score of 183 points), the South Korea team scored 38 points (with a total team score of 175 points), and the Thailand team scored 42 points (with a total team score of 174 points).

The gold medal cutoff for this IMO was set at 31 points (with 49 contestants earning gold medals), the silver medal cutoff was 24 points (with 112 contestants earning silver medals), and the bronze medal cutoff was 15 points (with 155 contestants earning bronze medals).

In this IMO, only one contestant achieved a perfect score of 42 points, namely Jinmin Li from China.

6.3 Summary

In the first 64 IMOs, there were a total of eight graph theory problems, as depicted in Figure 6.3. The score details for these problems are presented in Table 6.2. Due to the smaller number of participating teams and missing contestant score information in early IMOs, there are several blanks in Table 6.2.

These eight problems were proposed by seven countries. Hungary contributed two problems.

From Table 6.2, it can be observed that in the first 64 IMOs, there were two graph theory problems with an average score of 0–1 points; one problem with an average score of 1–2 points; one problem with an average score of 2–3 points; four problems with an average score of 3–4 points. Among these

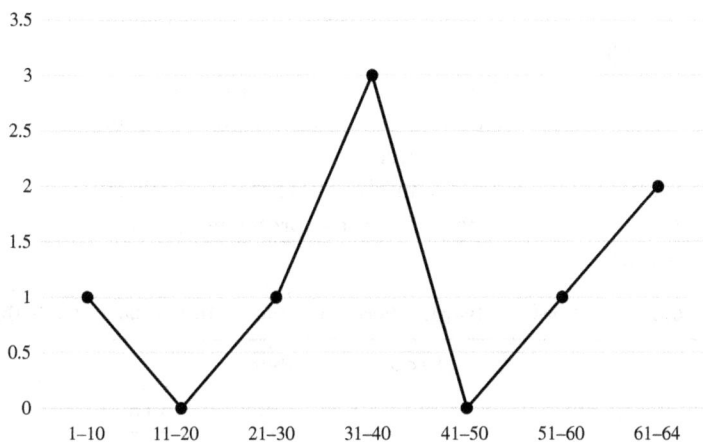

Figure 6.3 Numbers of Graph Theory Problems in the First 64 IMOs

Table 6.2 Score Details of Graph Theory Problems in the First 64 IMOs

Problem	6.1	6.2	6.3	6.4	6.5	6.6	6.7	6.8
Full points	6	7	7	7	7	7	7	7
Average score	2.471	3.084	3.542	3.128	1.903	0.572	0.940	3.938
Top five mean	2.025	4.950	6.133	4.833	5.367	3.367	3.700	6.867
6th–15th mean		3.250	4.848	5.183	3.633	1.833	2.083	6.700
16th–25th mean			4.056	3.183	1.700	0.981	1.583	6.283
Problem number in the IMO	6-4	21-2	31-2	32-4	33-3	60-3	61-3	61-4
Proposing country	Hungary	Bulgaria	Czechoslovakia	The United States	China	Croatia	Hungary	India

Note. Top five mean = Total score of the top five teams ÷ Total number of contestants from the top five teams,

6th–15th mean = Total score of the 6th–15th teams ÷ Total number of contestants from the 6th–15th teams,

16th–25th mean = Total score of the 16th–25th teams ÷ Total number of contestants from the 16th–25th teams.

problems, the one with the lowest average score is Problem 6.6 (IMO 60-3), proposed by Croatia.

In the 24th–64th IMOs, there were a total of six graph theory problems. Among these, two had an average score of 0–1 point; one had an average score of 1–2 points; three had an average score of 3–4 points. Further analysis of the problem numbers of these six graph theory problems, as shown in Table 6.3, reveals that these problems frequently appeared as the 3rd/6th problem.

Table 6.3 Numbers of Graph Theory Problems in the 24th–64th IMOs

Graph Theory Problem	Problem Number			Number of Problems in the First 64 IMOs
	1, 4	2, 5	3, 6	
Graph theory problems	2	1	3	8

From Table 6.2, it is observable that in the graph theory problems, the average score of the top five teams generally exceeds the average score of the problem by 3 points, the average score of the 6th–15th teams typically surpasses the average score by 1.5 points. Meanwhile, the average score of the 16th–25th teams is usually within 1 point of the average score of the problem.

Appendix A

IMO General Information

Session	Year	Host	Number of partici-pating teams	Number of con-testants	Gold Cutoffs/Numbers of Medalists	Silver	Bronze
IMO 1	1959	Romania	7	52	37/3	36/3	33/5
IMO 2	1960	Romania	5	39	40/4	37/4	33/4
IMO 3	1961	Hungary	6	48	37/3	34/4	30/4
IMO 4	1962	Czechoslovakia	7	56	41/4	34/12	29/15
IMO 5	1963	Poland	8	64	35/7	28/11	21/17
IMO 6	1964	The Union of Soviet Socialist Republics	9	72	38/7	31/9	27/19
IMO 7	1965	The German Democratic Republic	10	80	38/8	30/12	20/17
IMO 8	1966	Bulgaria	9	72	39/13	34/15	31/11
IMO 9	1967	Yugoslavia	13	99	38/11	30/14	22/26
IMO 10	1968	The Union of Soviet Socialist Republics	12	96	39/22	33/22	26/20
IMO 11	1969	Romania	14	112	40/3	30/20	24/21
IMO 12	1970	Hungary	14	112	37/7	30/11	19/40
IMO 13	1971	Czechoslovakia	15	115	35/7	23/12	11/29

(*Continued*)

(Continued)

Session	Year	Host	Number of participating teams	Number of contestants	Gold Silver Bronze (Cutoffs/Numbers of Medalists)		
IMO 14	1972	Poland	14	107	40/8	30/16	19/30
IMO 15	1973	The Union of Soviet Socialist Republics	16	125	35/5	27/15	17/48
IMO 16	1974	The German Democratic Republic	18	140	38/10	30/24	23/37
IMO 17	1975	Bulgaria	17	135	38/8	32/25	23/36
IMO 18	1976	Austria	18	139	34/9	23/28	15/45
IMO 19	1977	Yugoslavia	21	155	34/13	24/29	17/35
IMO 20	1978	Romania	17	132	35/5	27/20	22/38
IMO 21	1979	The United Kingdom	23	166	37/8	29/32	20/42
IMO 22	1981	The United States of America	27	185	41/36	34/37	26/30
IMO 23	1982	Hungary	30	119	37/10	30/20	21/31
IMO 24	1983	France	32	186	38/9	26/27	15/57
IMO 25	1984	Czechoslovakia	34	192	40/14	26/35	17/49
IMO 26	1985	Finland	38	209	34/14	22/35	15/52
IMO 27	1986	Poland	37	210	34/18	26/41	17/48
IMO 28	1987	Cuba	42	237	42/22	32/42	18/56
IMO 29	1988	Australia	49	268	32/17	23/48	14/65
IMO 30	1989	Germany	50	291	38/20	30/55	18/72
IMO 31	1990	The People's Republic of China	54	308	34/23	23/56	16/76
IMO 32	1991	Sweden	56	318	39/20	31/51	19/84
IMO 33	1992	The Russian Federation	56	322	32/26	24/55	14/74
IMO 34	1993	Turkey	73	413	30/35	20/66	11/97
IMO 35	1994	Chinese Hong Kong	69	385	40/30	30/64	19/98
IMO 36	1995	Canada	73	412	37/30	29/71	19/100
IMO 37	1996	India	75	424	28/35	20/66	12/99
IMO 38	1997	Argentina	82	460	35/39	25/70	15/122

(Continued)

(*Continued*)

Session	Year	Host	Number of participating teams	Number of contestants	Gold	Silver	Bronze
					Cutoffs/Numbers of Medalists		
IMO 39	1998	Chinese Taiwan	76	419	31/37	24/66	14/102
IMO 40	1999	Romania	81	450	28/38	19/70	12/118
IMO 41	2000	Republic of Korea	82	461	30/39	21/71	11/119
IMO 42	2001	The United States of America	83	473	30/39	20/81	11/122
IMO 43	2002	The United Kingdom	84	479	29/39	23/73	14/120
IMO 44	2003	Japan	82	457	29/37	19/69	13/104
IMO 45	2004	Greece	85	486	32/45	24/78	16/120
IMO 46	2005	Mexico	91	513	35/42	23/79	12/128
IMO 47	2006	Slovenia	90	498	28/42	19/89	15/122
IMO 48	2007	Vietnam	93	520	29/39	21/83	14/131
IMO 49	2008	Spain	97	535	31/47	22/100	15/120
IMO 50	2009	Germany	104	565	32/49	24/98	14/135
IMO 51	2010	Kazakhstan	95	522	27/47	21/103	15/115
IMO 52	2011	The Netherlands	101	563	28/54	22/90	16/137
IMO 53	2012	Argentina	100	547	28/51	21/88	14/137
IMO 54	2013	Colombia	97	527	31/45	24/92	15/141
IMO 55	2014	South Africa	101	560	29/49	22/113	16/133
IMO 56	2015	Thailand	104	577	26/39	19/100	14/143
IMO 57	2016	Chinese Hong Kong	109	602	29/44	22/101	16/135
IMO 58	2017	Brazil	111	615	25/48	19/90	16/153
IMO 59	2018	Romania	107	594	31/48	25/98	16/143
IMO 60	2019	The United Kingdom	112	621	31/52	24/94	17/156
IMO 61	2020	The Russian Federation	105	616	31/49	24/112	15/155
IMO 62	2021	The Russian Federation	107	619	24/52	19/103	12/148
IMO 63	2022	Norway	104	589	34/44	29/101	23/140
IMO 64	2023	Japan	112	618	32/54	25/90	18/170

Appendix B

IMO Combinatorics Problem Index

Problem number in the IMO	Proposing country or region	Category	Problem number in the book	Page number
IMO 5-6	Hungary	Logical Reasoning Problems, Operation and Logical Reasoning Problems	Problem 4.1	127
IMO 6-4	Hungary	Graph Theory Problems	Problem 6.1	230
IMO 6-5	Romania	Point and Line Problems, Combinatorial Geometry Problems	Problem 5.1	171
IMO 7-6	Poland	Point and Line Problems, Combinatorial Geometry Problems	Problem 5.2	172
IMO 8-1	The Soviet Union	Enumerative Combinatorics Problems	Problem 1.1	34
IMO 9-6	Hungary	Enumerative Combinatorics Problems	Problem 1.2	35

(*Continued*)

(*Continued*)

Problem number in the IMO	Proposing country or region	Category	Problem number in the book	Page number
IMO 11-5	Mongolia	Point and Line Problems, Combinatorial Geometry Problems	Problem 5.3	174
IMO 12-4	Czechoslovakia	Existence Problems	Problem 2.1	77
IMO 12-6	The Soviet Union	Shape Problems, Combinatorial Geometry Problems	Problem 5.18	198
IMO 13-5	Bulgaria	Point and Line Problems, Combinatorial Geometry Problems	Problem 5.4	175
IMO 13-6	Sweden	Enumerative Combinatorics Problems	Problem 1.3	37
IMO 14-1	The Soviet Union	Existence Problems	Problem 2.2	78
IMO 16-1	The United States	Logical Reasoning Problems, Operation and Logical Reasoning Problems	Problem 4.2	129
IMO 16-4	Bulgaria	Covering, Embedding, Partitioning, and Patching Problems, Combinatorial Geometry Problems	Problem 5.21	203
IMO 17-5	The Soviet Union	Point and Line Problems, Combinatorial Geometry Problems	Problem 5.5	176
IMO 18-4	The United States	Extremal Combinatorial Problems	Problem 3.1	111
IMO 19-2	Vietnam	Extremal Combinatorial Problems	Problem 3.2	113
IMO 20-6	The Netherlands	Existence Problems	Problem 2.3	79
IMO 21-2	Bulgaria	Graph Theory Problems	Problem 6.2	231
IMO 21-6	Germany	Enumerative Combinatorics Problems	Problem 1.4	38
IMO 22-2	Germany	Enumerative Combinatorics Problems	Problem 1.5	40
IMO 23-6	Vietnam	Point and Line Problems, Combinatorial Geometry Problems	Problem 5.6	177
IMO 24-4	Belgium	Point and Line Problems, Combinatorial Geometry Problems	Problem 5.7	179

(*Continued*)

(*Continued*)

Problem number in the IMO	Proposing country or region	Category	Problem number in the book	Page number
IMO 34-3	Finland	Single-Person Operation Problems, Operation and Logical Reasoning Problems	Problem 4.4	131
IMO 34-6	The Netherlands	Single-Person Operation Problems, Operation and Logical Reasoning Problems	Problem 4.5	134
IMO 36-3	Japan	Point and Line Problems, Combinatorial Geometry Problems	Problem 5.10	184
IMO 37-1	Finland	Single-Person Operation Problems, Operation and Logical Reasoning Problems	Problem 4.6	136
IMO 38-1	Belarus	Lattice Point and Grid Problems, Combinatorial Geometry Problems	Problem 5.24	211
IMO 38-4	Iran	Existence Problems	Problem 2.6	83
IMO 39-2	India	Enumerative Combinatorics Problems	Problem 1.12	52
IMO 40-1	Estonia	Point and Line Problems, Combinatorial Geometry Problems	Problem 5.11	185
IMO 40-3	Belarus	Extremal Combinatorial Problems	Problem 3.4	116
IMO 41-3	Belarus	Single-Person Operation Problems, Operation and Logical Reasoning Problems	Problem 4.7	138
IMO 41-4	Hungary	Single-Person Operation Problems, Operation and Logical Reasoning Problems	Problem 4.8	140
IMO 42-3	Germany	Existence Problems	Problem 2.7	85
IMO 43-1	Colombia	Enumerative Combinatorics Problems	Problem 1.13	53
IMO 44-1	Brazil	Enumerative Combinatorics Problems	Problem 1.14	56
IMO 45-3	Estonia	Covering, Embedding, Partitioning, and Patching Problems, Combinatorial Geometry Problems	Problem 5.22	205

(*Continued*)

(Continued)

Problem number in the IMO	Proposing country or region	Category	Problem number in the book	Page number
IMO 57-6	Czech	Point and Line Problems, Combinatorial Geometry Problems	Problem 5.16	195
IMO 58-3	Austria	Double-Person Operation Problems, Operation and Logical Reasoning Problems	Problem 4.14	152
IMO 58-5	Russia	Existence Problems	Problem 2.11	95
IMO 59-3	Iran	Existence Problems	Problem 2.12	96
IMO 59-4	Armenia	Double-Person Operation Problems, Operation and Logical Reasoning Problems	Problem 4.15	154
IMO 60-3	Croatia	Graph Theory Problems	Problem 6.6	236
IMO 60-5	The United States	Single-Person Operation Problems, Operation and Logical Reasoning Problems	Problem 4.10	144
IMO 61-3	Hungary	Graph Theory Problems	Problem 6.7	239
IMO 61-4	India	Graph Theory Problems	Problem 6.8	240
IMO 61-6	Chinese Taiwan	Point and Line Problems, Combinatorial Geometry Problems	Problem 5.17	196
IMO 62-5	Spain	Double-Person Operation Problems, Operation and Logical Reasoning Problems	Problem 4.16	156
IMO 63-1	Australia	Single-Person Operation Problems, Operation and Logical Reasoning Problems	Problem 4.11	146
IMO 63-6	Austria	Lattice Point and Grid Problems, Combinatorial Geometry Problems	Problem 5.27	218
IMO 64-5	The Netherlands	Existence Problems	Problem 2.13	99